Studies in Advanced and Selective Manufacturing Technologies

Studies in Advanced and Selective Manufacturing Technologies

Guest Editor

Muhannad Ahmed Obeidi

MDPI

Basel • Beijing • Wuhan • Barcelona • Belgrade • Novi Sad • Cluj • Manchester

Guest Editor
Muhannad Ahmed Obeidi
School of
Mechanical Engineering
Technological
University Dublin
Dublin
Ireland

Editorial Office
MDPI AG
Grosspeteranlage 5
4052 Basel, Switzerland

This is a reprint of the Special Issue, published open access by the journal *Designs* (ISSN 2411-9660), freely accessible at: https://www.mdpi.com/journal/designs/special_issues/12LZU6P48Z.

For citation purposes, cite each article independently as indicated on the article page online and as indicated below:

Lastname, A.A.; Lastname, B.B. Article Title. *Journal Name* **Year**, *Volume Number*, Page Range.

ISBN 978-3-7258-7795-9 (Hbk)
ISBN 978-3-7258-7796-6 (PDF)
https://doi.org/10.3390/books978-3-7258-7796-6

Contents

About the Editor

Muhannad Ahmed Obeidi

Muhannad Ahmed Obeidi is a Lecturer at Technological University Dublin, Ireland, where he delivers modules in manufacturing technologies, materials properties and processing, and engineering design to undergraduate and postgraduate students. His research and teaching focus on advanced manufacturing, particularly additive manufacturing, metal 3D printing, laser surface modification, and process optimisation. He has previously served as an Assistant Professor at Dublin City University and as a Funded Investigator at the I-Form Advanced Manufacturing Research Centre, where he led research on metal additive manufacturing and laser processing, integrating data-driven and Industry 4.0 approaches into both research and education.

Dr. Obeidi has extensive industrial experience as a Design Engineer at Nammo Ireland Ltd., contributing to the development of aerospace components, including rocket engine fuel control systems, with responsibilities spanning material selection, simulation, and validation under extreme conditions. His career also includes significant engineering and management roles in large-scale industrial projects and mechanical systems. He has authored over 80 scientific publications and a book on additive manufacturing process optimisation, and his work has gained international recognition, including being listed among the top 2% of scientists globally in a Stanford University and Elsevier study (2025).

Preface

This Reprint examines recent advances in advanced manufacturing technologies and their transformative role in engineering design. Its scope covers additive manufacturing, advanced material processing, architected materials, surface engineering, and selective machining, with a focus on how these approaches interact to influence material behaviour, structural performance, and functional efficiency. The primary aim is to demonstrate that manufacturing is no longer a secondary step, but a central design driver that integrates materials, geometry, and performance within a unified framework.

The motivation for this work arises from the growing need for lightweight, high-performance, and application-specific components across industries such as aerospace, energy, and transportation. Rapid technological progress—particularly in additive manufacturing—has created new opportunities for tailoring internal structures and material properties, while also introducing challenges related to process control, defects, and reliability. Addressing these complexities requires a holistic perspective that bridges design and manufacturing.

Bringing together nine peer-reviewed contributions, this volume provides insights into current research trends and practical developments. It is intended for researchers, engineers, and practitioners in manufacturing, materials science, and mechanical design who seek to understand and apply emerging technologies in advanced engineering systems.

Muhannad Ahmed Obeidi
Guest Editor

designs

MDPI

Editorial

Studies in Advanced and Selective Manufacturing Technologies

Muhannad Ahmed Obeidi

School of Mechanical Engineering, Technological University Dublin, Bolton Street, D01 K822 Dublin, Ireland; muhannad.ahmedobeidi@tudublin.ie

Advanced manufacturing technologies are increasingly shaping how engineering systems are designed, optimised, and realised. By enabling complex geometries, tailored material distributions, and controlled microstructural features, modern manufacturing processes have become essential drivers of innovation across sectors such as transportation, aerospace, energy, and tooling. This reprint book focuses on recent advances that illustrate how manufacturing is no longer a downstream activity, but a central design enabler that links materials, geometry, and performance within an integrated engineering framework.

The volume brings together nine peer-reviewed contributions published in Designs, offering a representative overview of current research trends in additive manufacturing, advanced material processing, lattice and cellular design, surface engineering, and selective machining technologies. Rather than addressing these topics in isolation, the collected works emphasise the interdependence between process parameters, material behaviour, and structural architecture in achieving lightweight, high-performance, and application-specific components.

A dominant theme throughout this collection is the role of additive manufacturing (AM) in performance-driven structural design. Recent studies have shown that AM enables unprecedented control over internal architectures and geometric features, allowing designers to tailor energy absorption, stiffness, and failure mechanisms through informed choices of infill strategy, wall thickness, curvature, and material system [1,2]. In both polymer- and metal-based AM, process optimisation has been shown to strongly influence defect formation, density, and microstructural evolution, which in turn govern mechanical response and functional reliability [3–5].

Beyond AM, advanced material processing routes continue to play a critical role in improving mechanical efficiency and multifunctionality. Powder-based fabrication strategies, including novel milling and consolidation approaches, have demonstrated significant potential for enhancing the strength–ductility balance and energy absorption capacity of lightweight composites [6,7]. Similarly, surface engineering and coating technologies remain essential for extending component lifespan and improving resistance to wear, oxidation, and mechanical degradation under demanding operating conditions [8].

Architected materials and structures form another key pillar of this reprint book. Lattice, cellular, and triply periodic minimal surface (TPMS) architectures are increasingly employed to achieve high stiffness-to-weight ratios, controlled deformation behaviour, and multifunctional performance. Prior studies have shown that the effective implementation of such structures depends not only on geometric design, but also on manufacturing constraints, load transfer mechanisms, and failure modes [9,10], reinforcing the need for close integration between design and process selection.

Finally, the continued importance of selective material removal processes is reflected in recent work on non-traditional machining techniques for advanced alloys. Several studies investigated processes such as abrasive water jet machining provide viable solutions for materials that are difficult to process using conventional methods, while still allowing acceptable surface integrity and dimensional accuracy when parameters are carefully controlled [11,12].

Overall, this reprint book demonstrates that the most impactful advances in advanced and selective manufacturing arise from a holistic consideration of materials, processes, and structural design. The collected contributions provide valuable insights for researchers and practitioners seeking to translate emerging manufacturing technologies into robust, efficient, and application-driven engineering solutions. It is hoped that this volume will serve both as a reference and as a catalyst for future research in advanced manufacturing and design.

Conflicts of Interest: The author declares no conflict of interest.

List of Contributions

1. Shahrjerdi, A.; Karamimoghadam, M.; Shahrjerdi, R.; Casalino, G.; Bodaghi, M. Optimizing Selective Laser Melting of Inconel 625 Superalloy through Statistical Analysis of Surface and Volumetric Defects. *Designs* **2024**, *8*, 87. https://doi.org/10.3390/designs8050087.
2. Basak, A.K.; Sali, J.M.; Pramanik, A. Mechanical Properties of 17-4 PH Stainless Steel Manufactured by Atomic Diffusion Additive Manufacturing. *Designs* **2025**, *9*, 66. https://doi.org/10.3390/designs9030066.
3. Rezvanpour, H.; Vergnano, A. A Flake Powder Metallurgy Approach for Fabricating Al/CNT Composites: Combining Dual-Matrix and Shift-Speed Ball Milling to Optimize Mechanical Properties. *Designs* **2025**, *9*, 82. https://doi.org/10.3390/designs9040082.
4. Basak, A.K.; Radhika, N.; Prakash, C.; Pramanik, A. Investigation on the Microstructure and Micro-Mechanical Properties of Thermal-Sprayed NiCoCrAlY High Entropy Alloy Coating. *Designs* **2024**, *8*, 37. https://doi.org/10.3390/designs8020037.
5. Freund, R.; Hilbig, K.; Vietor, T. A Systematic Evaluation of Design Freedoms and Restrictions of Lattice Structures Used as Interlock Bonds. *Designs* **2024**, *8*, 97. https://doi.org/10.3390/designs8050097.
6. Shahid, M.N.; Shahid, M.U.; Rasheed, S.; Irfan, M.; Obeidi, M.A. Computational Investigation of the Fluidic Properties of Triply Periodic Minimal Surface (TPMS) Structures in Tissue Engineering. *Designs* **2024**, *8*, 69. https://doi.org/10.3390/designs8040069.
7. Arumugam, A.; Pramanik, A.; Dixit, A.R.; Basak, A.K. Abrasive Water Jet Machining (AWJM) of Titanium Alloy—A Review. *Designs* **2026**, *10*, 13. https://doi.org/10.3390/designs10010013.
8. Saber, A.; Güler, M.A.; Acar, E.; ElSayed, O.S.; Aldallal, H.; Alsadi, A.; Aldousari, Y. Crash Performance of Additively Manufactured Tapered Tube Crash Boxes: Influence of Material and Geometric Parameters. *Designs* **2025**, *9*, 72. https://doi.org/10.3390/designs9030072.

References

1. Gibson, I.; Rosen, D.; Stucker, B. *Additive Manufacturing Technologies: 3D Printing, Rapid Prototyping, and Direct Digital Manufacturing*; Springer: New York, NY, USA, 2021.
2. Fahad, M.; Raja, W.; Iqbal, M.N.; Awan, A.W. Effect of Curvature Shape on the Impact Strength of Additively Manufactured Acrylonitrile Butadiene Styrene Parts Produced via Fused Deposition Modeling. *Designs* **2024**, *8*, 132. [CrossRef]
3. DebRoy, T.; Wei, H.L.; Zuback, J.S.; Mukherjee, T.; Elmer, J.W.; Milewski, J.O.; Beese, A.M.; Wilson-Heid, A.; De, A.; Zhang, W. Additive manufacturing of metallic components—Process, structure and properties. *Prog. Mater. Sci.* **2018**, *92*, 112–224. [CrossRef]
4. King, W.E.; Barth, H.D.; Castillo, V.M.; Gallegos, G.F.; Gibbs, J.W.; Hahn, D.E.; Kamath, C.; Rubenchik, A.M. Observation of keyhole-mode laser melting in laser powder-bed fusion additive manufacturing. *J. Mater. Process. Technol.* **2014**, *214*, 2915–2925. [CrossRef]
5. Lewandowski, J.J.; Seifi, M. Metal additive manufacturing: A review of mechanical properties. *Annu. Rev. Mater. Res.* **2016**, *46*, 151–186. [CrossRef]
6. Miracle, D.B. Metal matrix composites—From science to technological significance. *Compos. Sci. Technol.* **2005**, *65*, 2526–2540. [CrossRef]

7. Torralba, J.M.; da Costa, C.E.; Velasco, F. Powder metallurgy steel matrix composites: Properties and applications. *J. Mater. Process. Technol.* **2003**, *133*, 203–206. [CrossRef]
8. Pawlowski, L. *The Science and Engineering of Thermal Spray Coatings*, 2nd ed.; Wiley: Chichester, UK, 2008.
9. Deshpande, V.S.; Ashby, M.F.; Fleck, N.A. Foam topology: Bending versus stretching dominated architectures. *Acta Mater.* **2001**, *49*, 1035–1040. [CrossRef]
10. Maskery, I.; Aboulkhair, N.T.; Aremu, A.O.; Tuck, C.J.; Ashcroft, I.A.; Wildman, R.D.; Hague, R.J.M. A mechanical property evaluation of graded density Al-Si10-Mg lattice structures manufactured by selective laser melting. *Mater. Sci. Eng. A* **2017**, *670*, 264–274. [CrossRef]
11. Hashish, M. A modeling study of metal cutting with abrasive waterjets. *J. Eng. Mater. Technol.* **1984**, *106*, 88–100. [CrossRef]
12. Arumugam, A.; Pramanik, A.; Dixit, A.R.; Basak, A.K. Abrasive Water Jet Machining (AWJM) of Titanium Alloy—A Review. *Designs* **2026**, *10*, 13. [CrossRef]

https://doi.org/10.3390/designs10020024

designs

MDPI

Review

A Systematic Evaluation of Design Freedoms and Restrictions of Lattice Structures Used as Interlock Bonds

Raphael Freund, Karl Hilbig * and Thomas Vietor

Institute for Engineering Design, Technische Universität Braunschweig, Hermann-Blenk-Str. 42, 38108 Braunschweig, Germany; r.freund@tu-braunschweig.de (R.F.); t.vietor@tu-braunschweig.de (T.V.)
* Correspondence: k.hilbig@tu-braunschweig.de

Abstract: Additive manufacturing provides new possibilities in product design compared to traditional manufacturing processes. Particularly additive material extrusion offers the freedom to combine multiple materials in a single component without additional steps. However, combining multiple materials often leads to reduced adhesion, which can hinder the creation of high-strength designs. This issue can be largely mitigated using the geometric freedom of additive manufacturing to produce interlocking structures. This publication investigates the use of lattice structures as interlocking bonds in multi-material applications. The aim is to aid the design of suitable lattice structures by collecting geometric freedoms of lattices, application requirements, and manufacturing constraints, for this information to be used in suitable designs in the future. Initially, the general design freedoms of lattice structures are compiled and explained. Subsequently, these design freedoms are narrowed down based on the specific requirements for interlocking bonds and the limitations imposed by geometry and material combinations during manufacturing. The publication concludes with design recommendations that can be used as the basis for interlock bonds. Suitable lattice designs should aim for high interconnectivity, interconnected porosity, and a high number of similar strut structures, all the while maintaining low dimensions in the interface direction.

Keywords: additive manufacturing; multi-material; MEX; lattice structures; interlocks

1. Introduction

Additive manufacturing (AM) offers significant potential for product design, enabling new design freedoms and introducing new material possibilities, especially in multi-material manufacturing. Among the various AM technologies, additive material extrusion (MEX) stands out for its vast potential to produce multi-material components without additional process steps, thanks to the feasibility of combining different materials within a single layer. However, adhesion at the multi-material interface is typically low, posing challenges for product developers and manufacturers [1–3].

Despite this, the geometric and material flexibility of AM can often mitigate these issues. For example, in cases of poor adhesion, interlocking structures can be implemented to change the failure mode from adhesive failure at the interface to cohesive failure within the structure, significantly increasing the maximum load capacity [4–12]. This approach makes it easier to utilize material combinations with low adhesion. When using flexible materials, the requirements for such interlocks are more severe, as they tend to require three-dimensional interlocks to prevent extraction of the two interlock halves without breaking.

One example of well-suited 3D interlocks for this task is interlocking lattice structures. These structures have been shown to significantly increase bonding strength in multi-material applications or in adhesive bonds [7–9]. However, to fully utilize the potential of interlocking lattice structures for such bonding purposes, it is crucial to understand their design freedoms thoroughly, as their mechanical behavior is highly dependent on their geometric design. Therefore, this publication aims to examine and collect the various design freedoms of interlocking structures on a theoretical basis to aid their application in combining materials with poor adhesive properties into load-bearing structures. The aim is to narrow the vast design space of lattice structures down to a manageable size. This information may then be used to more effectively design interlocking structures for a larger spectrum of applications and broaden the prevalence and practical applications of multi-material AM.

Interlocking bonds can be used for different applications in which low adhesion poses a problem, as depicted in Figure 1. On the one hand, true multi-material additive manufacturing can produce such structures in a single process step. On the other hand, injecting material into the structure in a second step can result in a similar interlocking structure that has been shown to increase bonding strength in adhesive bonding [9]. This publication will mainly focus on the former option of combining two or more materials in a single 3D print, especially regarding combining soft and hard materials.

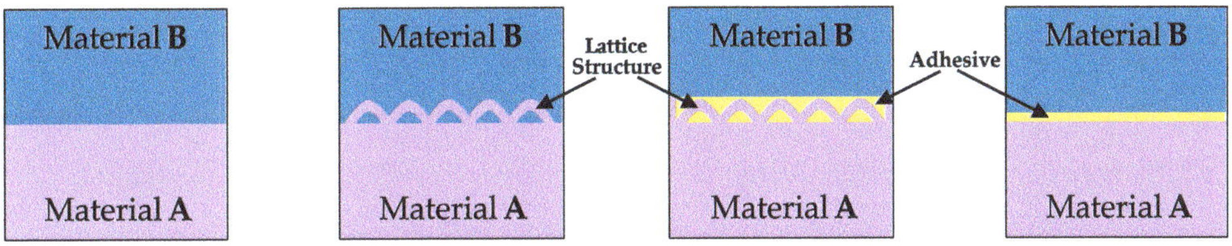

Figure 1. Schematic representation of different variants of multi-material composites using lattice structures. Left to right: simple multi-material AM-component, multi-material AM-component with interlocking bonds, interlock bonds with adhesive, simple adhesive bond [8].

2. Fundamentals and State of the Art

First, this section will briefly explain additive material as the used manufacturing process, as well as current applications of interlock bonds in MEX components.

2.1. Multi-Material Material Extrusion

Material extrusion (MEX, also known as FFF or FDM) is an additive manufacturing (AM) process that fabricates parts layer by layer by heating and plastifying thermoplastic polymer pre-product. Manufacturing costs of all AM processes are independent of the geometric complexity of the part, and mechanical properties are highly anisotropic due to the layer-by-layer process. By alternating between different extruders or by purging the nozzle with different materials during MEX fabrication, it is possible to combine various materials within the same part and even within the same layer. This capability allows for the manufacturing of multi-material parts without additional processing steps [13].

Using materials with varying properties such as hardness, temperature resistance, or electrical conductivity can be used as an effective means for function integration [14–37]. However, the mechanical properties of multi-material parts produced through this method depend significantly on the adhesive forces forming at the interface. These forces are influenced not only by the materials used but also by the process parameters during printing [38–47].

2.2. Interlocking Bonding Using Additive Manufacturing

Interlock bonding has been successfully utilized for increasing bonding strength in different applications [4–12]. These applications can be categorized into interlocks based on undercuts and complete three-dimensional interlocks. The former typically only shows undercuts in a single direction and can be separated if a single bonding partner deforms [4]. The latter, on the other hand, can never be separated without one partner breaking due to the two sides of the bond being topologically interlinked [7–9]. Three-dimensional interlocks are especially useful if the difference in the strength of the two bonding partners is high since the deformation of one partner is to be expected, which could cause undercut-based interlocks to separate.

Lattice structures are one example of such completely interlocking structures. They consist of a regular repeating pattern of unit cells made of plates or struts that typically offer a high performance-to-weight ratio [48,49] and high energy absorption [50,51]. They also provide a high degree of design freedom, enabling the creation of graded structures with finely tuned mechanical behavior [52,53]. Due to their high geometric complexity, these structures are exclusively manufactured using AM technologies. Due to the aforementioned interlocking, they hold significant potential for use in combination with multi-material AM in the form of mechanical locks.

By leveraging the design freedom of AM, these interlocking structures can achieve a high degree of complexity and can surpass conventional methods of enhancing adhesion, such as plasma treatment [9]. However, the behavior of these structures is not well understood, as they exhibit strong anisotropic behavior and often approach the resolution limits of manufacturing machines.

3. Influences on Composite Strength of Interlock Bonds

The composite strength of an interlocking bond primarily depends on the weakest part of the bond since the composite will fail if any part of the bond fails. As shown in Figure 2, four main factors contribute to the resulting composite strength, though only three can be influenced by design choices.

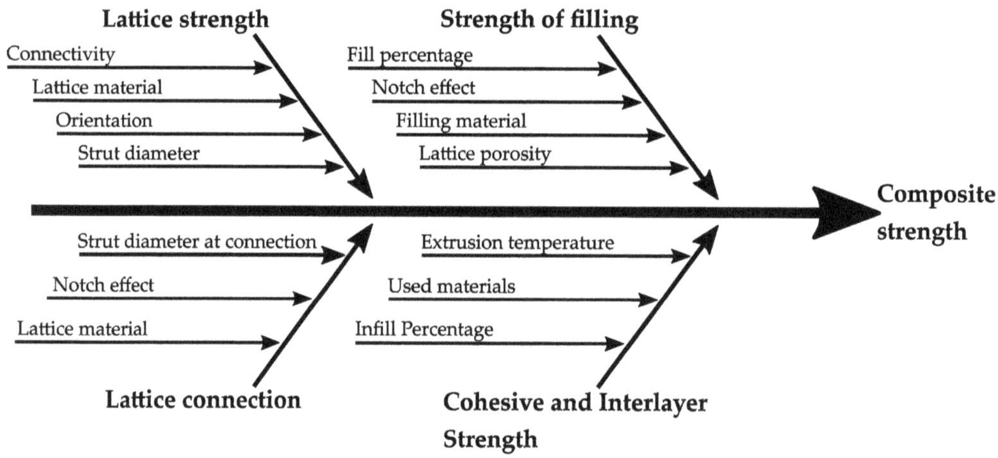

Figure 2. Influences on the composite strength of interlock bonds. Based on [8].

Starting from either side of the interlock bond, the cohesive strength in the areas outside of the interlock is determined by interlayer bonding. Failure outside of the bond cannot be meaningfully influenced by the lattice design but can be affected by proper process parameters that influence interlayer bonding, such as extrusion temperature. If failure occurs outside of the multi-material interface (i.e., in the area where only a single material is present), it indicates that the interlock bond withstood more force than the rest of the component could handle. This means that further improving the strength of the interlocking bond would not benefit this particular application since failure would once again occur at the weakest point of the component, i.e., the same position in the single-material area.

The next possible failure point is the fill of the lattice, which consists of a second material manufactured with MEX. Since the fill and the lattice intertwine in three dimensions, adhesive failure cannot cause separation of the interlocking structure without additional cohesive failure in the fill. While strong deformation can cause adhesive failure in the interlocking structures, forces can still be transmitted through the mechanical interlock in that case. Hence, to maximize composite strength, the fill's strength needs to be maximized by reducing stresses within it. This can be achieved by increasing the cross-sectional area, increasing the fill volume, reducing the notch effect at the interfaces between fill and lattice, and ensuring complete fill without defects or air bubbles.

Another possibility is failure that occurs in the lattice or its connection to the rest of the second substrate. Although lattice strength can be increased by increasing connectivity or rod diameter, this would decrease the fill's strength by reducing its occupied volume. Hence, a design conflict exists where the optimal composite strength results from balancing lattice strength with fill strength.

The last potential failure point is the connection between the lattice and the rest of the first substrate or correspondingly between the fill structure and the second structure. If the lattice strength, fill strength, and adhesive strength at the substrate are sufficiently high, the connections to the substrates will fail. This can be prevented by increasing the strut diameter at the connections (or correspondingly increasing the pore size at the top) or reducing the notch effect at these locations, again encountering the same design conflict as before.

It becomes clear that, apart from material combination, the design of the lattice is the most important factor contributing to the strength of an interlocking bond. To design structures well suited for the application, it is essential to have knowledge of all design levers and apply them appropriately.

4. Design Freedoms of Interlocking Lattices

To make an informed decision regarding geometries that can be used for interlock design, it is essential to systematically evaluate the design freedoms offered by lattice structures first. This evaluation is conducted for all possible lattice structures in the following section. The identified design freedoms of lattice structures are subsequently refined in later sections, considering manufacturability in relation to MEX and the practicality of interlocking bond applications.

4.1. Cell Structure

When evaluating the use of lattice structures as interlock bonds, it is important to first consider the advantages of different types of cell structures. While technically, only periodically repeating structures form 'real' lattices, stochastic and quasiperiodic structures share many properties with periodic lattices and hence should be considered alongside them.

The geometric freedoms of periodic lattice structures are best assessed based on the Bravais lattices based on crystallography. They describe all possible arrangements of points that can be repeated infinitely in 3-dimensional space. When connected to a lattice graph, these points create all possible arrangements of strut-based lattice structures. Each lattice mathematically consists of periodic lattice points, which are described by $t(u, v, w) = u \times a + v \times b + w \times c$ with u, v, w being integers and a, b, c, being the basis vectors of the lattice, repeat infinitely. All possible lattices can be categorized into 14 Bravais lattices according to their symmetry, which in turn can be sorted into 7 lattice systems. Each lattice point in these lattices is identical with regards to symmetry to each other lattice point. Hence, the addition of face- or body-centers restricts the possible geometry of the lattice, as shown in Figure 3 [54,55].

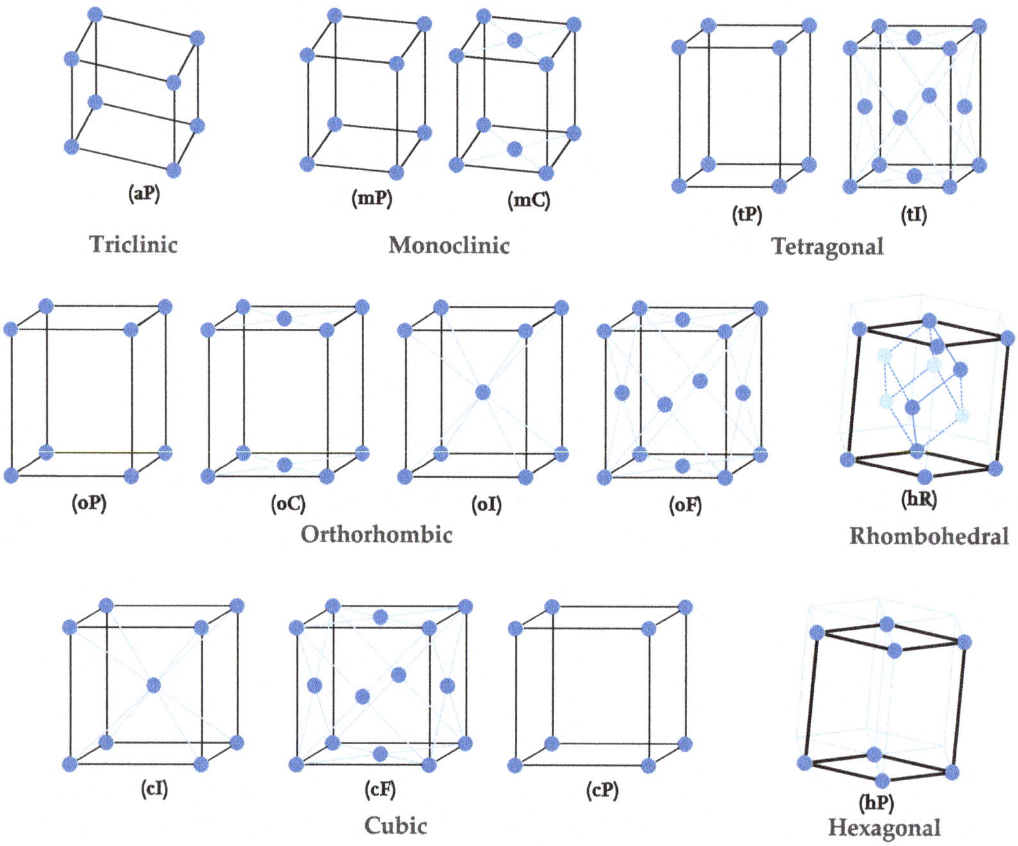

Figure 3. Bravais lattices and lattice systems ((**aP**): triclinic, (**cI**): cubic body-centered, (**cF**): cubic face-centered, (**cP**): cubic primitive, (**mP**): monoclinic primitive, (**mC**): monoclinic C-centered, (**oP**): orthorhombic primitive, (**oC**): orthorhombic C-centered, (**oI**): orthorhombic body-centered, (**oF**): orthorhombic face-centered, (**tP**): tetragonal primitive, (**tI**): tetragonal body-centered, (**hP**): hexagonal primitive, (**hR**): rhombohedral). Based on [54,56].

To create physical lattice structures from those mathematical lattice types, the cells must be populated with structural components based on the same or a higher symmetrical class. Besides symmetry, the behavior of lattice structures depends heavily on the geometry of the repeated unit cell. To maintain lattice symmetry, this cell geometry must follow the same symmetrical restrictions as the corresponding lattice type, lest the symmetry collapses into a less symmetrical Bravais lattice. Due to the periodic nature of lattices, the origin point can be chosen arbitrarily; however, to simplify the discourse, we will assume that the cell geometry goes through the origin. Due to symmetry, each other lattice point within the unit cell must be part of the cell geometry as well.

Furthermore, to create only a single lattice, we will assume that the geometry of the unit cell is one connected volume structure. This means that once the cell geometry is repeated ad infinitum, all other lattice points are also part of the emerging space-filling structure. The same holds for equivalent cell edges or cell faces–all equivalent points remain equivalent no matter their placement in the lattice. For example, if a point on one face of a cubic face-centered lattice Figure 3(cF) is part of the structure, all points at the same spot on all cubic faces as well as all corners also must be part of the lattice and are completely indistinguishable from each other. Additionally, since all structures inside of the cell must be connected to the lattice structure in some way, they must also be connected to the origin. This means that no matter the lattice type, all corner points are connected to the lattice, and all points inside a unit cell are connected.

4.2. Cell Type

Based on these restrictions, three different types of lattice structures are possible. If the emerging lattice graph is populated by linear struts as interconnections, the forming lattice will be called "1D-based", while "2D-based" lattices form if the interconnection is implemented through shell structures. More complex structures based on volumes will be called "3D-based". Examples of these three types are depicted in Figure 4.

Figure 4. Different types of unit cells. (**a**) 1D-based cell, (**b**) 2D-based cell, (**c**) volume-based cell [8].

Applications of these types are radically different. Since the emerging lattice based on these structures behaves differently, a possible use case must be taken into account when choosing between the three. One-dimensional-based lattices have large areas of interconnected empty space. In addition, their interconnectivity can easily be tailored, which affects mechanical behaviors like rigidity. Two-dimensional-based lattices can form more than one separate area of interconnected empty space or many non-connected pores. However, the structure is usually strongly interconnected, which makes tailoring the mechanical behavior more difficult. Three-dimensional-based lattices have the least design restrictions and, hence, can form a wide variety of structures with hugely different behaviors. Hence, no simple statement can be made in regard to these structures. However, many cases of 3D structures can show similar designs as 1D or 2D-based cells. The structure shown in Figure 4c, for example, can either be thought of as a cube with a sphere-shaped hole or as a cubic strut-based lattice with peculiar node and strut shapes, and hence is similar to a 1D-based cell. Therefore, the structure depicted in Figure 4c. is expected to behave similarly to a simple cubic lattice consisting of cylindrical struts due to similar interconnectivity and pore configuration.

Due to the nature of their unit cells not being filled, lattice structures inherently exhibit some degree of porosity. In an ideal lattice structure devoid of boundary effects, the porosity of both the lattice structure and the unit cell are identical, being solely determined by the unit cell's geometry. This porosity can vary based on the elements constituting the unit cell. For example, line-based unit cells typically exhibit open porosity in all directions (except in cases where lines fuse due to their thickness). Conversely, shell and volume-based cells can display a range from fully closed to fully open pores, as illustrated in Figure 5.

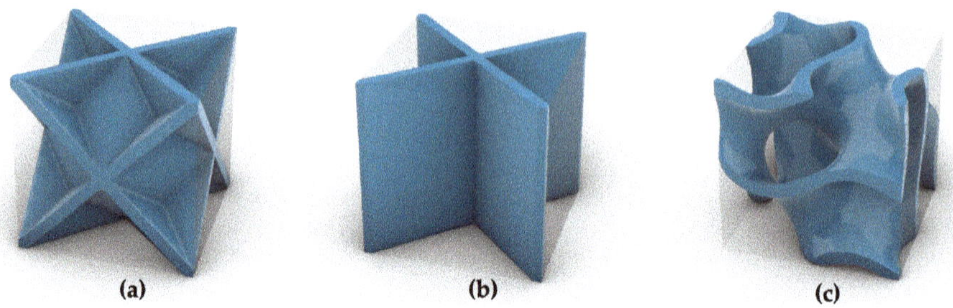

Figure 5. Different shell-based unit cells. (a) closed porous cell, (b) open porous cell in one direction, and (c) fully interconnected open porous cell in all directions [8].

Unit cells that resemble extrusion elements, such as honeycombs or the structure shown in Figure 5b, generally exhibit porosity in only one direction. When these extrusion-based elements are oriented differently but remain unconnected, the result is multiple unidirectional porosities that are also not interconnected. By interconnecting these one-dimensional structures, it is possible to achieve porosities in one, two, or multiple directions with shell or volume-based cells.

The porosity of the structure, whether open or closed, is crucial for many applications as it influences the mechanical behavior and stress distribution within the lattice structure, as well as its permeability. This permeability is particularly important for applications requiring infiltration, and hence, open porosity is a key requirement for creating interlock bonds.

4.3. Structural Rigidity

The mechanical behavior of strut-based lattice structures is largely influenced by their interconnectivity. The stiffness of these structures is determined by the number of connections (struts) and points (nodes), which can be calculated using the Maxwell criterion. This criterion is represented by the equation [57,58]:

$$M = s - 3n + 6$$

where s stands for the number of struts and n represents the number of nodes.

If the Maxwell number (M) is less than zero, it indicates that the system of struts has more degrees of freedom than constraints, resulting in flexible or under-stiff behavior. Conversely, if M equals zero, the structure has an equal number of constraints and degrees of freedom, making it just-stiff, with rigid behavior and no internal stresses. When M is greater than zero, the network is considered over-determined and behaves in an over-stiff manner.

Under-stiff structures show high compliance and tend to be weak, exhibiting bending-dominated behavior due to their excess degrees of freedom. However, since interlock bonding demands high stiffness and low weight, under-stiff lattices cannot be used. Furthermore, the higher the strength-to-weight ratio, the less volume the lattice takes up within the bonding zone. Consequently, more volume is taken up by the interlocking fill structure, increasing its strength and, therefore, the strength of the whole bond. Just-stiff and over-stiff structures, on the other hand, have very low compliance and high strength because of their limited degrees of freedom, resulting in stretch-dominated behavior. Over-stiff structures, in particular, further reduce compliance due to strut redundancy, making them desirable for many lattice structure applications.

When determining their strength regarding the Maxwell criterion, lattices always need to be considered whole, as neighboring cells usually have a stiffening effect on each other. Hence, a single unit cell may have a Maxwell number less than zero and thus be considered under-stiff, even though the structure may turn out to be over-stiff once the criterion is applied to the periodic lattice as a whole. For instance, a cubic unit cell with eight corner nodes and the corresponding edge struts, along with a single body diagonal strut, as shown in Figure 6, may appear under-stiff with a Maxwell number of −5.

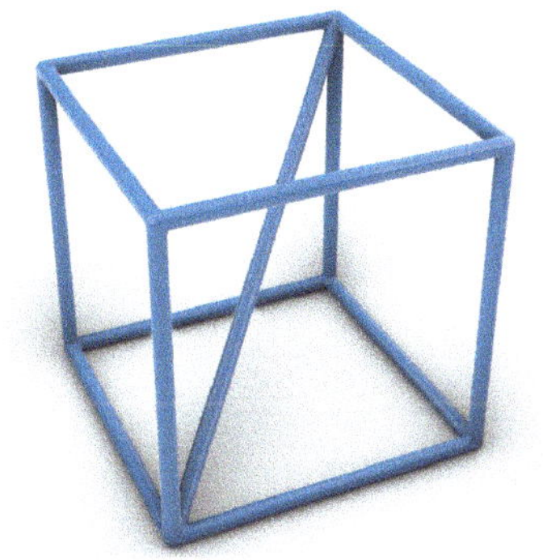

Figure 6. Single cell with a Maxwell number of −5. Based on [8].

However, when considering that each node and strut is part of multiple unit cells within the lattice structure, the mechanical properties become reliant on the quantity and configuration of these unit cells within the lattice. In an infinite arrangement, each corner node is part of eight distinct unit cells, while each edge is part of four unit cells. Similarly, structures located on the faces of the unit cell are shared between two adjacent unit cells, whereas those contained within the cell are not shared at all. Through normalizing, the number of struts and nodes can be adjusted to new values, denoted as s_0 and n_0, representing the effective count of struts and nodes within each cell, considering all surrounding cells.

Therefore, for a significantly large number of unit cells (N), the Maxwell criterion can be modified to:

$$M = N \times s_0 - N \times 3n_0 + 6$$

To determine whether a structure is over-stiff or under-stiff, only the sign of the Maxwell number (positive or negative) is relevant. Hence, both sides of the equation can be divided by N. As the number of unit cells increases, the constant term approaches zero, and consequently, the equation simplifies to (with $M' = M/N$):

$$M' = s_0 - 3n_0$$

Using this equation, sufficiently large lattices can be categorized by only considering the unit cell. This way, seemingly under-stiff unit cells can be shown to combine into an over-stiff lattice structure when considering the interconnection and arrangement of multiple unit cells.

4.4. Orientation

To enhance stiffness, not only the number and types of struts but also their orientation and placement are crucial factors. Struts exhibit greater resistance to tensile and compressive loads than to bending, making it advantageous to orient them such that the angle between the struts and the load direction is minimized. The simplest method to achieve this is by rotating the lattice, although other strategies are also viable. In cubic lattices, all edges form right angles with each other, while face diagonals and body diagonals form angles of 45° and 35.3° with the cubic faces, respectively. By distorting the unit cell, these directions can be adjusted to better suit specific applications, as the selection of unit vectors determines the angles and orientation.

Based on the Bravais lattices, the least restricted unit cell shape without right angles or equal-length sides is a parallelepiped. By aligning the three unit vectors with the primary load directions, the displacement of the struts can be minimized. Additionally, parallelepiped cells can be individually distorted, provided their corners still align with neighboring unit cells. Although the resulting structure technically no longer qualifies as a lattice (due to the loss of long-range order), it retains many lattice-like characteristics due to its repeating nature. An example of this distorted structure is illustrated in Figure 7c.

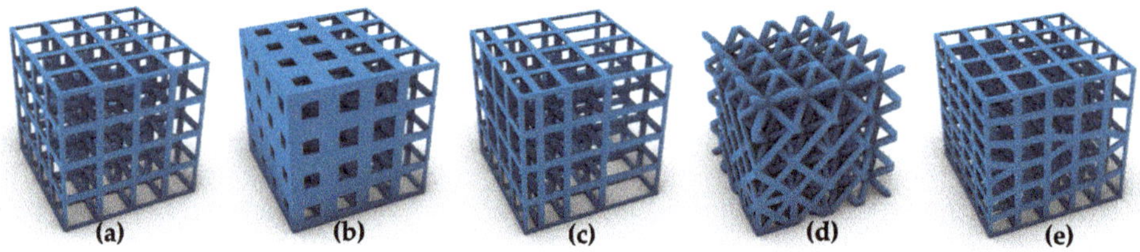

Figure 7. Different types of density gradients. (**a**) no gradient, (**b**) strut diameter gradient, (**c**) cell size gradient, (**d**) gradient through displacements, (**e**) gradient through strut addition. Based on [8].

4.5. Density Gradients

To enhance lattice stiffness, controlling the density along the load path is crucial, as stiffness and toughness are primarily influenced by density. Several methods exist to modify cell density: altering strut or shell diameters, adjusting cell size, introducing displacements, or changing the unit cell by modifying the number of struts. These density gradient methods are illustrated in Figure 7.

4.5.1. Strut Diameter and Shell Thickness

The simplest way to adjust lattice density is by changing the strut or shell diameter, as shown in Figure 7b. For cells with low density, where overlapping volume is negligible, lattice density is proportional to shell thickness for shell-based structures and the square of the strut diameter for line-based structures. As overlap increases, the relationship becomes more complex, with the density approaching 100%.

4.5.2. Cell Size

Density can also be adjusted by changing the cell dimensions while keeping the strut or shell thickness constant, as shown in Figure 7c. If parallelepiped cell angles remain constant, cells in each row along the x-vector must have the same y- and z-dimensions. For large, non-overlapping cells, density is inversely proportional to the cell dimension or, equivalently, to the cube of the side length if all side lengths scale equally. When cell size decreases and overlap becomes significant, the density behavior changes, approaching a fixed value based on lattice geometry. This relationship results in a linear correlation when cell size is expressed as the number of cells in a fixed volume. When cell dimensions change equally in all directions, the relationship for strut-based cells becomes quadratic.

As explained above, distorting cell dimensions also changes the angles within the unit cell and might, therefore, affect mechanical behavior in an unintended way.

4.5.3. Unit Cell Modification

Lattice density can also be adjusted by adding or removing struts or shells from the unit cell, as can be seen in Figure 7d. The resulting density varies based on element type, orientation, and thickness. In low-density cells with minimal overlap, the volume fraction is roughly proportional to the sum of strut lengths or shell areas. While the lower bound for density is reached once no more struts can be removed without disconnecting cell corners, density can still be increased by adding shell elements or curved lines even when all connections between lattice points are present. Although modifying the unit cell in this way is less efficient for density adjustment than changing diameter or cell size, it can be beneficial for specific applications, such as reinforcing areas under load, without significantly increasing overall structure weight.

4.5.4. Displacements

Lattice structures can incorporate dislocations similar to real-world crystals, as shown in Figure 7e. Adding half-planes with non-parallelepiped cells at their edges allows for incremental changes in the number of cells within a given volume. However, these wedge-shaped added cells must fit neighboring cells, causing surrounding cells to warp to maintain spatial continuity. This once again changes the angles within the distorted cells. Adding displacements increases cell number while dimensions stay constant, similarly affecting density as described for cell size changes.

4.6. Combination of Methods or Lattices

All methods of altering density are independent and can be applied simultaneously, offering a vast design space for using lattice structures in technical applications. Optimizing a component's density distribution can be implemented using one or more of these methods.

While this publication focuses on using lattice structures as a basis for interlock bonds, other applications of lattice structures, especially when combined with gradients, are also possible. Gradients allow for fine-tuning of mechanical properties, enabling adjustments to structures for various load cases. By combining several load-specific gradients in different directions, multiple load cases can be considered during the design process, as illustrated below.

Furthermore, different lattices can be combined using Boolean operations while ensuring proper orientation of the relevant struts, as described above. Each partial lattice can then have gradients applied to it separately, guaranteeing proper load transmission for each specific load case, as depicted in Figure 8.

Figure 8. Lattice with a cell size gradient in the z-direction and two different unit cells, each with a different strut diameter gradient. (Body diagonals have a gradient in the y-direction, and edges have a gradient in the x-direction). Based on [8].

5. Restrictions of Interlocking Lattices Manufactured with MEX

The following section will narrow down the design freedoms stated above based on the specific requirements for interlocking bonds and the limitations imposed by geometry and material combinations during manufacturing. The goal is to arrive at a set of rules that, when met, should provide a good fit for the design of interlock bonds.

5.1. Restrictions in Regard to Function

The implementation of lattice structures as interlock bonds imposes several design constraints, primarily stemming from the required mechanical behavior of the bonds.

To minimize the volume needed for the interlock, the strength of the bonds should be maximized, with energy absorption being a secondary consideration. This necessitates a specific design of lattice structures in regard to the freedoms discussed above.

5.1.1. Cell Structure and Orientation

To minimize the volume taken up by the interlock bond, the cell structure should ideally have minimal height. Preferably, a single layer of cells or even half-cells should be used so that the volume occupied by the bond is minimal. As the interlock forms in this very first layer, it is assumed that every additional layer has a decreasing contribution to the bonding strength or even reduces strength by increasing the number of possible failure points. The latter is expected to be especially relevant when the two interlocking materials have very similar strengths.

The symmetry of the cells should ideally follow the symmetry of the load cases. Depending on the load type, different symmetries can be reasonable. Since tensile or compression loads put almost equal stress over the whole cross-sectional area of the bond, a bond optimized for such loads should show a high level of symmetry in the xy-direction and avoid areas of stress concentration. Shear loads in adhesive bonds, on the other hand, form a parabolic stress distribution in the z-direction with a peak at the overlap [59]. However, the case presents more complex interlock bonds since the paths the forces take are likely influenced by the geometry of the lattice. Lastly, peel loads will always apply a line load to the lattice. To avoid load peaks, ideally the lattice should avoid sudden geometric changes perpendicular to the line direction. This is impossible, however, as each strut or wall constitutes such a change. Hence, a steady number of small changes should be preferred. This means a larger number of struts and a lattice with no unit vectors along the line load direction are desirable.

To increase strut count while maintaining porosity, the cells should be as small as possible in the xy-direction. Subsequently, the forces transmitted through each strut are lower, and failure of a single strut has less effect, reducing the chance of cascading failure.

5.1.2. Cell Type

Ideally, unit cells for interlock bonds should exhibit interconnected porosity in three dimensions, as this is the only way to form non-separable undercut areas of two materials with highly different Young's moduli. Line-based unit cells are particularly advantageous for interlock bonds because they are straightforward to design and predict, and they form three-dimensional interlocks in almost all cases.

Furthermore, the bonds should possess high strength in multiple directions and load cases, indicating the need for a symmetric design. Additionally, eliminating certain failure modes is crucial to enhancing the bonding strength to the greatest extent possible. Uniformity of stresses is the most crucial factor, as failure of only a single strut leads to overburdening of the surrounding structures and cascading failure. Hence, the lattice should be as uniform as possible to distribute stresses as evenly as possible. This means that different types of struts should be avoided, and the lattice should consist of only a single type of strut (horizontal edges, vertical edges, face diagonals, or body diagonals). This also means that, in most cases, gradients are not worthwhile.

5.1.3. Dimensioning and Pore Size

In an interlock bond, the weaker partner determines the overall bond strength. Moreover, enhancing the lattice strength by adjusting its dimensions typically reduces the strength of the interlock partner, and vice versa, as the available volume changes. Consequently, lattice dimensions should be selected in such a way as to ensure that its strength matches that of the interlocking structure, maximizing the overall bond strength. Since the strength of both partners is influenced by the mechanical properties of the materials used, no straightforward conclusion can be drawn without knowledge of the materials. Therefore, the final choice of dimensions must be validated through experiments or simulations. Since the interlock forms through pores inside the lattice, pore size is a crucial factor, especially in cases where the interlock forms not through layered manufacturing but through injection. In cases of high-viscosity materials being injected, larger pores are expected to have a positive impact, as they likely can achieve more complete filling.

5.1.4. Structural Rigidity

Complex stress states, and anisotropic behavior within the lattice should be avoided, as they result in stress concentration in certain struts, which can lead to cascading failures when the critical stress of one strut type is exceeded, therefore overloading the remaining struts. Therefore, it is advantageous to include only a single type of strut in the interlock bond if possible.

For interlock bonds, stretch-dominated behavior is a desirable trait since the strength of the bending-dominated lattice is lower. Structural rigidity criteria dictate that the behavior of lattices is significantly influenced by their interconnectivity and the strut-to-node ratio. Using only a single layer of cells (explanation above) inherently results in low interconnectivity, making the Maxwell criterion the only pertinent guideline for designing rigid interlocking structures. Given that bonding areas are typically extensive, requiring lattice structures to encompass a substantial number of cells, the constant term in Maxwell's equation becomes negligible. Thus, a strut-to-node ratio of at least 3:1 should be maintained.

To achieve this, it is necessary to prioritize certain strut types over others, as the overall number of struts must be minimized to achieve large pore sizes, and certain strut directions yield higher connectivity. Consequently, only strut types with high connectivity should be selected. Among all possible options, edge struts exhibit the lowest strut-to-vertex ratio of 2:1 and should, therefore, be excluded from interlock bonding structures. Furthermore, the anisotropic behavior of edge struts can lead to premature failure under load, resulting in cascading failures that must be avoided. Most importantly, vertical struts show no undercut and, therefore, are useless for creating interlocks. Hence, their inclusion does not sufficiently enhance structural strength to a degree that warrants the reduced space available for the interlock partner.

Similarly, face diagonal struts should also be avoided, though with slightly less urgency, for analogous reasons. Including a single diagonal into the face of the cubic structure increases the number of struts while not affecting the number of nodes, therefore increasing interconnectivity; however, adding second diagonals into the same cubic face requires an additional node at the face center and does not further increase lattice connectivity while introducing additional failure points. While a cubic structure with no diagonals has a normalized Maxwell number of $M' = 0$ (i.e., behaves just-stiff when the structure is infinite), a cubic structure with a single diagonal on each face has a normalized Maxwell number of $M' = 3$ (i.e., is stiff) while a structure with two diagonals on each cubic face also has a normalized Maxwell number of $M' = 3$. Although horizontal face diagonals provide a large undercut area, they typically require a second row of vertices, which contradicts the requirement of a single layer of struts. Body diagonals exhibit the highest connectivity, making them the preferred choice for interlock bonds. Including a single body diagonal once again adds struts without requiring additional nodes, increasing interconnectivity and making it a safe choice. Furthermore, adding more body diagonals into the structure never decreases interconnectivity, even if an additional node is required. A structure with a single body diagonal has a normalized Maxwell number of $M' = 1$. Structures with two, three, or four body diagonals result in values of $M' = 1$, $M' = 3$, and $M' = 5$, respectively, making all of them over-stiff. Experimental results demonstrate that structures composed exclusively of body diagonals exhibit excellent mechanical performance and provide sufficient undercut area to ensure robust interlock bond strength [60]. Although specific use cases may require additional struts, body diagonals should be prioritized and included in almost every lattice designed for interlock bonding applications.

5.1.5. Gradients

While gradients exhibit significant potential for tailoring the mechanical behavior of lattice structures to specific load conditions, interlock bonds are not expected to benefit from density gradients. In these scenarios, density gradients cause local weakening in one portion of the interlock while simultaneously strengthening the other. Because the composite strength is determined by the weakest element of the interlock, the presence of density gradients ultimately results in a reduction in overall composite strength for these types of loads. In most cases, the formation of cracks in the weakened portion seems to be the limiting factor and uniform stress distribution overall shows better results [7,8].

5.1.6. Lattice Base and Top

If the lattice and filling are strong enough, failure will move out from the lattice to the base or the top. To avoid failure due to crack formation at the base of the lattice, all struts of walls should be well-rounded to avoid a notch effect. Similarly, the top of the lattice can lead to crack formation in the interlock partner. Hence, the top of the lattice should show curvature as well (typically at nodes connecting several struts). This is especially important, as notches will occur between layers in any case due to layer-by-layer manufacturing.

5.2. Manufacturing Restrictions in Regards to Geometry

Restrictions regarding the manufacturability of lattice structures are dependent on the utilized manufacturing process and manufacturing machine. These restrictions can predominantly be expressed as dimensional restrictions that are multiples of line width or layer height. The academic literature provides extensive guidance regarding design for additive manufacturing. Due to the relationship between manufacturing machine and design restrictions it is impossible to exhaustively list all restrictions. Hence, the following is not a comprehensive list but provides a brief overview of relevant design restrictions for lattice structures manufactured by MEX. Interested readers should refer to relevant literature for detailed design rules [61–67].

5.2.1. External Radius of Single-Curved Elements

The external radius should be large enough so that each layer of a component is built up from a contour with an inscribed grid ($r_a \geq 1.5\,\text{mm}$) [61] (rule 12).

5.2.2. Internal Radius of Single-Curved Elements

The internal radius should be sufficiently large to prevent the inner sides of the elements from closing ($r_i \geq 0.3\,\text{mm}$) [61] (rule 13).

5.2.3. Length of Single-Curved Elements

The length of unsupported single-curved elements should be short enough to prevent deformation or excessive vibration. The critical length depends on the overhang angle. ($l \leq 10 \times r_a$ (for $\alpha = 45°$) or $l \leq 20 \times r_a$ (for $\alpha = 90°$)) [61] (rule 13).

5.2.4. Firmly Bonded Element Transitions

Edges should be rounded with a radius corresponding to the outer/inner radius of simply curved elements ($r_k = r_a \geq 1.5\,\text{mm}$ or $r_k = r_i \geq 0.3\,\text{mmrk} = r_i \geq 0.3\,\text{mm}$) [61] (rule 30).

5.2.5. Surface Angles

The surface angle of downward-facing surfaces should be large enough to avoid the need for solid support structures ($\alpha \geq 35°$) [61] (rule 52).

5.3. Manufacturing Restrictions in Regards to Material

Different materials exhibit different demands in regard to a variety of process parameters. These different requirements further restrict the possible designs of multi-material lattice structures. Once again, a non-comprehensive list shall provide an overview of relevant material-dependent restrictions.

5.3.1. Build Chamber or Build Plate Temperature

To reduce warping some materials are usually processed using a heated build plate or build chamber. This might pose a problem in multi-material prints when the required build chamber temperature of one material exceeds the heat deflection temperature or even the decomposition temperature of the second material. In such a case, designs need to take into account that lowering the build chamber temperature can increase warping. Hence, designs need to be more resilient against thermal stresses, e.g., by increasing corner radii, etc.

Similar problems might arise when different build plate temperatures are required for the different materials. However, this restriction is typically confined to only the first couple of layers that experience the temperature of the build plate.

5.3.2. Interface Adhesion

Two materials used in the same MEX process can form two different types of interfaces based on the orientation of the interface.

Vertical interfaces form when different materials are used in the same layer. Since lines of molten material are usually placed onto an existing layer (or onto the build plate in the case of the first layer) they mainly stick to already structures consisting of the same material located underneath. Since adhesion is high in such cases, the moving nozzle does not exert enough force to pull the extruded structure off, and consequently, it stays in place.

However, if two different materials are used in consecutive layers right on top of one another, the adhesion at the material interface becomes crucial. The interface can only be manufactured if adhesion is strong enough to keep the extruded structures in place. Even if geometric structures like mechanical interlocks can stop adhesive failure from occurring, such structures only form through the addition of more than one layer and hence serve no benefit in the interface layer. While additional measures are possible, like plasma treatment or the inclusion of a layer consisting of a third material to promote adhesion, they require special equipment (i.e., a plasma nozzle integrated into the manufacturing process or special printers capable of using three or more materials). In practice, these solutions are rarely applicable, and in most cases, insufficient adhesion results in an incompatible material combination.

As a result, the minimal requirement for the formation of interlock bonds is adhesion sufficient to survive the manufacturing process for at least a couple of layers. Once the interlock has formed, forces are mainly transmitted via mechanical interlocking, and adhesion becomes irrelevant.

6. Recommendation for Lattice Design for Interlock Bonding

Based on the restrictions listed above, factors for good lattice design can be stated, and the design space of lattices for mechanical interlock bonding can be reduced to a manageable size.

Since the number of different strut types is to be reduced, connectivity should be as high as possible to achieve stretch-dominated behavior. Body diagonal struts are well suited for this task, but other strut types are applicable as well, as long as the interconnectivity of the resulting lattice is high.

The number of struts in z-direction should be minimal to reduce the number of failure points. Hence, ideally, a single layer of struts is used. The base and top of this lattice should be well-rounded to avoid a notch effect and crack formation.

To produce equal behavior in all directions, each strut should show a similar angle. Hence, a lattice consisting of only a single type of strut with a side ratio close to 1:1 seems like a good option. Minor adjustments might be necessary to avoid incomplete cells (and, therefore, differently behaving struts).

The dimensioning of such lattices highly depends on the manufacturing restrictions resulting from the utilized 3D printer. These restrictions mainly influence cell size and strut diameter.

While smaller cell sizes are beneficial due to the higher number of struts, in the case of MEX, the strut diameter should be larger than 6 times the line width to maintain manufacturability. A typical line width is 0.4 mm, resulting in a minimal strut diameter of 2.4 mm.

While the height of the structure should be kept as low as possible, all strut directions must exceed the minimal overhang angle in order to minimize the low adhesion overlap of different materials. Assuming body diagonal struts are mainly used in the lattice, the ratio of cell height to cell width is important, as it determines the angle of all non-edge struts. The relationship between the overhang angle and the full cell dimensions x, y, and z of body diagonals is given through $tan\alpha = \frac{z}{\sqrt{x^2+y^2}}$. For example, assuming a quadratic cell base and a critical overhang angle of 30° (a typical value for MEX machines) results in a minimal cell height of 0.82 times the cell width to ensure manufacturability, see Figure 9.

Figure 9. Example of a lattice design based on the listed restrictions. A single layer of half-cells filled with a body-centered cubic lattice structure. Based on [8].

As mentioned above, the exact dimensioning of such structures depends on the load case and the materials used. Since there are different possibilities to manufacture such lattice structures and form an interlock, the number of possible material combinations is quite high.

Manufacturing of these structures can be accomplished with a variety of AM processes. While the manufacturing restrictions in this publication focus on MEX, adjusting the restrictions according to other processes like powder bed fusion or vat polymerization is easily possible, even while true multi-material manufacturing is significantly more difficult with these processes. This only matters, however, if the interlock is to be formed by multi-material AM.

If an adhesive or a second material is injected into the lattice structure, where it hardens [9,68], producing such structures is by no means restricted to AM processes with multi-material capability.

7. Conclusions and Future Prospects

Additive manufacturing, particularly through additive material extrusion, offers unprecedented flexibility in product design by enabling the integration of multiple materials into a single component without requiring additional steps. Even though interlocking lattice structures can address the challenge of reduced adhesion between different materials, such geometries are rarely used for this purpose. This may be partially due to the large design space, which makes creating well-suited structures difficult.

For this reason, this study has collected the geometric freedoms inherent in lattice structures and additive manufacturing, taking into account the specific requirements and constraints of multi-material applications. The approach involved narrowing down the general design freedoms of lattice structures to address the unique challenges of interlocking bonds and the limitations imposed by both geometric considerations and material combinations during the manufacturing process.

The result is a small number of guidelines that can aid the design process of interlocking lattices as interfaces in multi-material structures and provide a foundational framework for the design of suitable lattice structures in multi-material additive manufacturing.

Suitable lattice designs should have large pores interconnected in three dimensions, high lattice interconnectivity, and a high number of similar structures while maintaining low dimensions in the interface direction. While this still allows for a large number of designs to be used in the application of interlock bonds, these restrictions severely restrict the solution space of possible designs and can help future designers produce suitable geometries. In strut-based lattices, a single layer of cubic body diagonal struts can meet these restrictions, for example.

While the exact design of such structures depends on the restrictions of the used manufacturing machine as well as the moduli and strengths of the used materials, conventional design guidelines for MEX and geometric considerations of the lattice (e.g., porosity vs. modulus-ratio, strut angle vs. maximum overhang angle, strut diameter vs. extrusion width, etc.) can be used to further restrict the design space and generate suitable geometries.

Hence, this work can aid future research and practical applications by enabling designers to leverage the full potential of additive manufacturing to create robust, high-performance multi-material components.

Author Contributions: Conceptualization, R.F. and K.H.; methodology, R.F.; validation, R.F.; formal analysis, R.F.; investigation, R.F.; resources, T.V. and K.H.; data curation, R.F.; writing—original draft preparation, R.F.; writing—review and editing, R.F. and K.H.; visualization, R.F.; supervision, T.V.; project administration, T.V.; funding acquisition, T.V. and R.F. All authors have read and agreed to the published version of the manuscript.

Funding: Funded by the German Research Foundation (DFG)—Methods for design of additively manufactured multi-material parts and characterization of bonding strength (467577506) and Universal design for adaptive handles for adjustment to specific operating scenarios (452009430). We acknowledge the financial support from German Research Foundation and Technische Universität Braunschweig.

Data Availability Statement: Data are contained within the article.

Conflicts of Interest: The authors declare no conflicts of interest.

References

1. Jabbari, E.; Peppas, N.A. Polymer-Polymer Interdiffusion and Adhesion. *J. Macromol. Sci. Part C Polym. Rev.* **1994**, *34*, 205–241. [CrossRef]
2. Nazir, A.; Gokcekaya, O.; Md Masum Billah, K.; Ertugrul, O.; Jiang, J.; Sun, J.; Hussain, S. Multi-material additive manufacturing: A systematic review of design, properties, applications, challenges, and 3D printing of materials and cellular metamaterials. *Mater. Des.* **2023**, *226*, 111661. [CrossRef]
3. Vaezi, M.; Chianrabutra, S.; Mellor, B.; Yang, S. Multiple material additive manufacturing—Part 1: A review. *Virtual Phys. Prototyp.* **2013**, *8*, 19–50. [CrossRef]
4. Leuteritz, G.; Schudak, S.; Rohling, M.; Lachmayer, R. Formschlüssige multimaterielle Additive Fertigung zur Realisierung funktionsintegrierter Hybrid-Bauteile. In *Konstruktion für die Additive Fertigung 2018*; Lachmayer, R., Lippert, R.B., Kaierle, S., Eds.; Springer: Berlin/Heidelberg, Germany, 2020; pp. 299–317. ISBN 978-3-662-59057-7.
5. Peralta Marino, G.; de La Pierre, S.; Salvo, M.; Díaz Lantada, A.; Ferraris, M. Modelling, additive layer manufacturing and testing of interlocking structures for joined components. *Sci. Rep.* **2022**, *12*, 2526. [CrossRef] [PubMed]
6. Verma, S.; Yang, C.-K.; Lin, C.-H.; Jeng, J.Y. Additive manufacturing of lattice structures for high strength mechanical interlocking of metal and resin during injection molding. *Addit. Manuf.* **2022**, *49*, 102463. [CrossRef]
7. Freund, R.; Sallach, F.M.; Vietor, T. Design Freedoms of Lattice Structures for Interlock Bonding. In *Future Automotive Production Conference 2022*; Dröder, K., Vietor, T., Eds.; Springer: Wiesbaden, Germany, 2023; pp. 156–169. ISBN 978-3-658-39927-6.
8. Freund, R. Lattice Structures as Interface Design of Additively Manufactured Composite Components. Ph.D. Dissertation, Technische Universität Braunschweig, Braunschweig, Germany, 2023.

9. Freund, R.; Koch, S.; Watschke, H.; Stammen, E.; Vietor, T.; Dilger, K. Utilization of additively manufactured lattice structures for increasing adhesive bonding using material extrusion. *J. Adhes.* **2021**, *100*, 340–361. [CrossRef]
10. Ni, Y.; Bai, H.; Wang, Z.; Liao, H.; Wu, W. Bio-inspired, metal additive manufacturing interlocked structures: Geometrically design and fracture performance analysis. *Compos. Struct.* **2023**, *321*, 117220. [CrossRef]
11. Alsheghri, A.A.; Alageel, O.; Amine Mezour, M.; Sun, B.; Yue, S.; Tamimi, F.; Song, J. Bio-inspired and optimized interlocking features for strengthening metal/polymer interfaces in additively manufactured prostheses. *Acta Biomater.* **2018**, *80*, 425–434. [CrossRef]
12. Englert, L.; Heuer, A.; Engelskirchen, M.K.; Frölich, F.; Dietrich, S.; Liebig, W.V.; Kärger, L.; Schulze, V. Hybrid material additive manufacturing: Interlocking interfaces for fused filament fabrication on laser powder bed fusion substrates. *Virtual Phys. Prototyp.* **2022**, *17*, 508–527. [CrossRef]
13. Gibson, I.; Rosen, D.; Stucker, B. *Additive Manufacturing Technologies*; Springer: New York, NY, USA, 2015; ISBN 978-1-4939-2112-6.
14. Andriya, N.; Dutta, V.; Vani, V.V. Study on 3D printed auxetic structurebased non-pneumatic tyres (NPT'S). *Mater. Manuf. Process.* **2022**, *37*, 1280–1297. [CrossRef]
15. Espalin, D.; Alberto Ramirez, J.; Medina, F.; Wicker, R. Multi-material, multi-technology FDM: Exploring build process variations. *Rapid Prototyp. J.* **2014**, *20*, 236–244. [CrossRef]
16. Deng, Y.; Wang, Z.; Shen, H.; Gong, J.; Xiao, Z. A comprehensive review on non-pneumatic tyre research. *Mater. Des.* **2023**, *227*, 111742. [CrossRef]
17. Fang, G.; Matte, C.D.; Kwok, T.H.; Wang, C.C. Geometry-based Direct Simulation for Multi-Material Soft Robots. In Proceedings of the 2018 IEEE International Conference on Robotics and Automation (ICRA), Brisbane, QLD, Australia, 21–25 May 2018.
18. Fischer, A.; Gebauer, S.; Khavkin, E. *3D-Druck im Unternehmen. Entscheidungsmodelle, Best Practices und Anwendungsbeispiele. Am Beispiel Fused Layer Modeling (FLM)*; Hanser: München, Germany, 2018.
19. Georgopoulou, A.; Vanderborght, B.; Clemens, F. Fabrication of a Soft Robotic Gripper with Integrated Strain Sensing Elements Using Multi-Material Additive Manufacturing. *Front. Robot. AI* **2021**, *8*, 615991. [CrossRef] [PubMed]
20. Ion, A.; Frohnhofen, J.; Wall, L.; Kovacs, R.; Alistar, M.; Lindsay, J.; Lopes, P.; Chen, H.-T.; Baudisch, P. Metamaterial Mechanisms. In Proceedings of the 29th Annual Symposium on User Interface Software and Technology, Tokyo, Japan, 16–19 October 2016.
21. Käfer, M.; Dohnal, F.; Goettgens, V.; Stajkovic, J.; Brunner, M.; Leichtfried, G. Experimental verification of additively manufactured stacked multi-wedge acoustic black holes in beams for low frequency. *Mech. Syst. Signal Process.* **2024**, *208*, 111065. [CrossRef]
22. Dul, S.; Fambri, L.; Pegoretti, A. Filaments Production and Fused Deposition Modelling of ABS/Carbon Nanotubes Composites. *Nanomaterials* **2018**, *8*, 49. [CrossRef] [PubMed]
23. Hampel, B.; Monshausen, S.; Schilling, M. Properties and applications of electrically conductive thermoplastics for additive manufacturing of sensors. *tm-Tech. Mess.* **2017**, *84*, 593–599. [CrossRef]
24. Gackowski, B.M.; Sharma, M.; Idapalapati, S. Multi-material additive manufacturing of self-heating structures for out-of-autoclave post-processing and de-icing. *Addit. Manuf.* **2023**, *68*, 103519. [CrossRef]
25. Jayanth, N.; Senthil, P. Application of 3D printed ABS based conductive carbon black composite sensor in void fraction measurement. *Compos. Part B Eng.* **2019**, *159*, 224–230.
26. Kim, K.; Park, J.; Suh, J.H.; Kim, M.; Jeong, Y.; Park, I. 3D printing of multiaxial force sensors using carbon nanotube (CNT)/thermoplastic polyurethane (TPU) filaments. *Sens. Actuators A Phys.* **2017**, *263*, 493–500. [CrossRef]
27. Kwok, S.W.; Goh, K.H.H.; Tan, Z.D.; Tan, S.T.M.; Tjiu, W.W.; Soh, J.Y.; Ng, Z.J.G.; Chan, Y.Z.; Hui, H.K.; Goh, K.E.J. Electrically conductive filament for 3D-printed circuits and sensors. *Appl. Mater. Today* **2017**, *9*, 167–175. [CrossRef]
28. Leigh, S.J.; Bradley, R.J.; Purssell, C.P.; Billson, D.R.; Hutchins, D.A. A simple, low-cost conductive composite material for 3D printing of electronic sensors. *PLoS ONE* **2012**, *7*, e49365. [CrossRef] [PubMed]
29. Reyes, C.; Somogyi, R.; Niu, S.; Cruz, M.A.; Yang, F.; Catenacci, M.J.; Rhodes, C.P.; Wiley, B.J. Three-Dimensional Printing of a Complete Lithium Ion Battery with Fused Filament Fabrication. *ACS Appl. Energy Mater.* **2018**, *1*, 5268–5279. [CrossRef]
30. Rymansaib, Z.; Iravani, P.; Emslie, E.; Medvidović-Kosanović, M.; Sak-Bosnar, M.; Verdejo, R.; Marken, F. All-Polystyrene 3D-Printed Electrochemical Device with Embedded Carbon Nanofiber-Graphite-Polystyrene Composite Conductor. *Electroanalysis* **2016**, *28*, 1517–1523. [CrossRef]
31. Yan, Y.; Jiang, Y.; Ng, E.L.L.; Zhang, Y.; Owh, C.; Wang, F.; Song, Q.; Feng, T.; Zhang, B.; Li, P.; et al. Progress and opportunities in additive manufacturing of electrically conductive polymer composites. *Mater. Today Adv.* **2023**, *17*, 100333. [CrossRef]
32. Watschke, H.; Hilbig, K.; Vietor, T. Design and Characterization of Electrically Conductive Structures Additively Manufactured by Material Extrusion. *Appl. Sci.* **2019**, *9*, 779. [CrossRef]
33. Watschke, H.; Goutier, M.; Heubach, J.; Vietor, T.; Leichsenring, K.; Böl, M. Novel Resistive Sensor Design Utilizing the Geometric Freedom of Additive Manufacturing. *Appl. Sci.* **2020**, *11*, 113. [CrossRef]
34. Zhang, D.; Chi, B.; Li, B.; Gao, Z.; Du, Y.; Guo, J.; Wei, J. Fabrication of highly conductive graphene flexible circuits by 3D printing. *Synth. Met.* **2016**, *217*, 79–86. [CrossRef]
35. Zhuang, Y.; Song, W.; Ning, G.; Sun, X.; Sun, Z.; Xu, G.; Zhang, B.; Chen, Y.; Tao, S. 3D–printing of materials with anisotropic heat distribution using conductive polylactic acid composites. *Mater. Des.* **2017**, *126*, 135–140. [CrossRef]
36. Ehrmann, G.; Blachowicz, T.; Ehrmann, A. Magnetic 3D-Printed Composites-Production and Applications. *Polymers* **2022**, *14*, 3895. [CrossRef] [PubMed]

37. von Petersdorff-Campen, K.; Hauswirth, Y.; Carpenter, J.; Hagmann, A.; Boës, S.; Schmid Daners, M.; Penner, D.; Meboldt, M. 3D Printing of Functional Assemblies with Integrated Polymer-Bonded Magnets Demonstrated with a Prototype of a Rotary Blood Pump. *Appl. Sci.* **2018**, *8*, 1275. [CrossRef]
38. Ermolai, V.; Sover, A.; Nagîț, G. Influence of bond interface over the lap-shear performance of 3D printed multi-material samples. *MATEC Web Conf.* **2022**, *368*, 1005. [CrossRef]
39. Rabbi, M.F.; Chalivendra, V. Improvement in interfacial fracture toughness of multi-material additively manufactured composites through thermal annealing. *Forces Mech.* **2021**, *5*, 100051. [CrossRef]
40. Rabbi, M.F.; Chalivendra, V. Interfacial fracture characterization of multi-material additively manufactured polymer composites. *Compos. Part C Open Access* **2021**, *5*, 100145. [CrossRef]
41. Delia, S.; Rochman, A.; Curmi, A. Factors affecting interface bonding in multi-material additive manufacturing. *Prog. Addit. Manuf.* **2024**, *9*, 1365–1379. [CrossRef]
42. Goh, G.L.; Lee, S.; Cheng, S.H.; Goh, D.J.S.; Laya, P.; van Nguyen, P.; Han, B.S.; Yeong, W.Y. Enhancing interlaminar adhesion in multi-material 3D printing: A study of conductive PLA and TPU interfaces through fused filament fabrication. *MSAM* **2024**, *3*, 2672. [CrossRef]
43. Freund, R.; Watschke, H.; Heubach, J.; Vietor, T. Determination of Influencing Factors on Interface Strength of Additively Manufactured Multi-Material Parts by Material Extrusion. *Appl. Sci.* **2019**, *9*, 1782. [CrossRef]
44. Baca, D.; Ahmad, R. The impact on the mechanical properties of multi-material polymers fabricated with a single mixing nozzle and multi-nozzle systems via fused deposition modeling. *Int. J. Adv. Manuf. Technol.* **2020**, *106*, 4509–4520. [CrossRef]
45. Baca Lopez, D.M.; Ahmad, R. Tensile Mechanical Behaviour of Multi-Polymer Sandwich Structures via Fused Deposition Modelling. *Polymers* **2020**, *12*, 651. [CrossRef]
46. Yin, J.; Lu, C.; Fu, J.; Huang, Y.; Zheng, Y. Interfacial bonding during multi-material fused deposition modeling (FDM) process due to inter-molecular diffusion. *Mater. Des.* **2018**, *150*, 104–112. [CrossRef]
47. Lopes, L.R.; Silva, A.F.; Carneiro, O.S. Multi-material 3D printing: The relevance of materials affinity on the boundary interface performance. *Addit. Manuf.* **2018**, *23*, 45–52. [CrossRef]
48. Posser, T.; Freitas de Oliveira, B. Design for additive manufacturing applied for mass reduction of a two-stroke engine cylinder for portable machine. *Int. J. Interact. Des. Manuf.* **2020**, *14*, 709–717. [CrossRef]
49. Reyes Belmonte, M.A.; Copeland, C.D.; Hislop, D.; Hopkins, G.; Schmieder, A.; Bredda, S.; Akehurst, S. Improving Heat Transfer and Reducing Mass in a Gasoline Piston Using Additive Manufacturing. In *SAE 2015 World Congress & Exhibition, APR. 21, 2015*; SAE Technical Paper Series; SAE International400 Commonwealth Drive: Warrendale, PA, USA, 2015.
50. Evans, A.G.; He, M.Y.; Deshpande, V.S.; Hutchinson, J.W.; Jacobsen, A.J.; Carter, W.B. Concepts for enhanced energy absorption using hollow micro-lattices. *Int. J. Impact Eng.* **2010**, *37*, 947–959. [CrossRef]
51. Pingle, S.M.; Fleck, N.A.; Deshpande, V.S.; Wadley, H.N.G. Collapse mechanism maps for a hollow pyramidal lattice. *Proc. R. Soc. A* **2011**, *467*, 985–1011. [CrossRef]
52. Maskery, I.; Hussey, A.; Panesar, A.; Aremu, A.; Tuck, C.; Ashcroft, I.; Hague, R. An investigation into reinforced and functionally graded lattice structures. *J. Cell. Plast.* **2017**, *53*, 151–165. [CrossRef]
53. Wang, Y.; Zhang, L.; Daynes, S.; Zhang, H.; Feih, S.; Wang, M.Y. Design of graded lattice structure with optimized mesostructures for additive manufacturing. *Mater. Des.* **2018**, *142*, 114–123. [CrossRef]
54. Julian, M.M. *Foundations of Crystallography with Computer Applications*, 2nd ed.; CRC Press: Hoboken, NJ, USA, 2015; ISBN 978-1-4665-5292-0.
55. Sands, D.E. *Introduction to Crystallography*; Dover Publisher: Mineola, NY, USA, 1993; ISBN 978-0486678399.
56. Napy1kenobi. Bravais Lattices. Available online: https://commons.wikimedia.org/wiki/File:Bravais_lattices.svg (accessed on 15 July 2024).
57. Maxwell, J.C.L. On the calculation of the equilibrium and stiffness of frames. *Lond. Edinb. Dublin Philos. Mag. J. Sci.* **1864**, *27*, 294–299. [CrossRef]
58. Deshpande, V.S.; Ashby, M.F.; Fleck, N.A. Foam topology: Bending versus stretching dominated architectures. *Acta Mater.* **2001**, *49*, 1035–1040. [CrossRef]
59. Bednarcyk, B.A.; Aboudi, J.; Yarrington, P.W. *Determination of the Shear Stress Distribution in a Laminate from the Applied Shear Resultant—A Simplified Shear Solution*; NASA STI: Hanover, MD, USA, 2007. [CrossRef]
60. Li, P.; Yang, F.; Bian, Y.; Zhang, S.; Wang, L. Designable mechanical properties of modified body-centered cubic lattice materials. *Compos. Struct.* **2023**, *317*, 117060. [CrossRef]
61. Adam, G.A.O. Systematische Erarbeitung von Konstruktionsregeln Für Die Additiven Fertigungsverfahren Lasersintern, Laserschmelzen und Fused Deposition Modeling. Ph.D. Dissertation, University Paderborn, Paderborn, Germany, 2015. ISBN 978-3-8440-3474-5.
62. Hallmann, M.; Schleich, B.; Wartzack, S. Erarbeitung von Gestaltungsrichtlinien für die Konstruktion additiv gefertigter Mechanismen. In *Konstruktion für die Additive Fertigung 2018*; Lachmayer, R., Lippert, R.B., Kaierle, S., Eds.; Springer: Berlin/Heidelberg, Germany, 2020; pp. 129–143. ISBN 978-3-662-59057-7.
63. Rudolph, J.-P. Prüfung auf Fertigungsrestriktionen und Konstruktionsrichtlinien. In *Cloudbasierte Potentialerschließung in der Additiven Fertigung*; Rudolph, J.-P., Ed.; Springer: Berlin/Heidelberg, Germany, 2018; pp. 53–80. ISBN 978-3-662-58262-6.

64. Junk, S.; Bär, F. Design guidelines for Additive Manufacturing using Masked Stereolithography mSLA. *Procedia CIRP* **2023**, *119*, 1122–1127. [CrossRef]
65. Emmelmann, C.; Herzog, D.; Kranz, J. Design for laser additive manufacturing. In *Laser Additive Manufacturing*; Elsevier: Amsterdam, The Netherlands, 2017; pp. 259–279. ISBN 9780081004333.
66. Gibson, I.; Rosen, D.; Stucker, B.; Khorasani, M. (Eds.) Design for Additive Manufacturing. In *Additive Manufacturing Technologies*; Springer International Publishing: Cham, Switzerland, 2021; pp. 555–607. ISBN 978-3-030-56126-0.
67. Diegel, O.; Nordin, A.; Motte, D. *A Practical Guide to Design for Additive Manufacturing*; Springer: Singapore, 2019; ISBN 978-981-13-8280-2.
68. Kazmer, D.O.; Colon, A. Injection printing: Additive molding via shell material extrusion and filling. *Addit. Manuf.* **2020**, *36*, 101469. [CrossRef]

designs

MDPI

Review

Abrasive Water Jet Machining (AWJM) of Titanium Alloy—A Review

Aravinthan Arumugam [1], Alokesh Pramanik [2], Amit Rai Dixit [3] and Animesh Kumar Basak [4,5,*]

1 School of Mechanical Engineering, Engineering Institute of Technology, Perth, WA 6005, Australia; aravin.arumugam@eit.edu.au
2 School of Civil and Mechanical Engineering, Curtin University, Perth, WA 6102, Australia; alokesh.pramanik@curtin.edu.au
3 Department of Mechanical Engineering, Indian Institute of Technology (Indian School of Mines), Dhanbad 826004, India; amitraidixit@iitism.ac.in
4 Adelaide Microscopy, Adelaide University, Adelaide, SA 5005, Australia
5 Centre for Research Impact & Outcome, Chitkara University Institute of Engineering and Technology, Chitkara University, Rajpura 140401, India
* Correspondence: animesh.basak@adelaide.edu.au

Abstract

Abrasive water jet machining (AWJM) is a non-traditional machining process that is increasingly employed for shaping hard-to-machine materials, particularly titanium (Ti)-based alloys such as Ti-6Al-4V. Owing to its non-thermal nature, AWJM enables effective material removal while minimising metallurgical damage and preserving subsurface integrity. The process performance is governed by several interacting parameters, including jet pressure, abrasive type and flow rate, nozzle traverse speed, stand-off distance, jet incident angle, and nozzle design. These parameters collectively influence key output responses such as the material removal rate (MRR), surface roughness, kerf geometry, and subsurface quality. The existing studies consistently report that the jet pressure and abrasive flow rate are directly proportional to MRR, whereas the nozzle traverse speed and stand-off distance exhibit inverse relationships. Nozzle geometry plays a critical role in jet acceleration and abrasive entrainment through the Venturi effect, thereby affecting the cutting efficiency and surface finish. Optimisation studies based on the design of the experiments identify jet pressure and traverse speed as the most significant parameters controlling the surface quality in the AWJM of titanium alloys. Recent research demonstrates the effectiveness of artificial neural networks (ANNs) for process modelling and optimisation of AWJM of Ti-6Al-4V, achieving high predictive accuracy with limited experimental data. This review highlights research gaps in artificial intelligence-based fatigue behaviour prediction, computational fluid dynamics analysis of nozzle wear mechanisms and jet behaviour, and the development of hybrid AWJM systems for enhanced machining performance.

Keywords: abrasive water jet machining (AWJM); titanium alloy; abrasive particles; material removal rate; machine learning

1. Introduction

Titanium alloys are mixtures of pure titanium and other metals, such as aluminium, chromium, cobalt, copper, iron, manganese, vanadium, and molybdenum, to impart titanium with the refined physical properties preferred in many industrial applications.

Titanium (Ti) alloys can be grouped as "α-alloys, near α-alloys, α + β alloys, near β-alloys and β-alloys," based on the crystalline structure and alloying element [1]. The most commonly used Ti-alloy, Ti-6Al-4V, goes to the "α + β alloy" classification and covers almost 50% usage of Ti alloys in manufacturing [2]. Titanium alloys possess a "high strength-to-weight ratio, high strength to stiffness at elevated temperatures, superior corrosion resistance and fatigue resistance and can sustain moderate to high temperatures without creeping" [3], which categorises them as "hard-to-machine" materials. Table 1 gives the mechanical properties of the Ti-6Al-4V alloy.

Table 1. Common properties of Ti-6Al-4V alloy, adapted from [4].

Properties	Values
Knoop hardness	362 HRC
Hardness	349 HV
Ultimate tensile strength	0.95 GPa
Elastic modulus	112 GPa
Density	4.4 g/cc
Poisson's ratio	0.35
Elongation	14%

Components are manufactured by machining materials to the required end shape, size, and surface requirements. Metals and alloys are machined using conventional machining through plastic shearing, hence leading to the "formation of a chip (for example, broaching, drilling, milling, turning), abrasion (for example, lapping, grinding) or micro-chipping (for example, polishing micro-abrasive and blasting)". Having said that, it is not recommended to apply "conventional machining processes to hard-to-machine materials, such as metal matrix composites, nickel alloys and titanium alloys such as Ti-6Al-4V" [5–8]. This is because of issues such as high "tool wear", poor "surface finish", longer machining time, and the inability to attain complex/intricate design and shapes. In addition, the relatively lower elastic modulus, low rate of thermal conductivity, and higher chemical affinity at an elevated temperature upsurge the cost related to the machining of Ti-alloy [9,10]. The "Metallurgical distinctiveness of titanium alloys makes it complicated and costlier to machine than that of steel with a similar range of hardness" [11].

Figure 1 depicts an evaluation of the conventional machinability of a number of titanium alloys and where it stands against other materials. Despite their "poor machinability, titanium alloys are widely used in medical (for example, dental implants, spinal fusion cages, finger and toe replacements and bone plate expandable ribs cages), aerospace, marine, chemical processing and automobile sectors" [12]. The depth of milling has a non-linear relationship with milling time for different materials, and for materials such as titanium, which has a "low machinability index, the non-linearity is more prominent" [13].

The present manuscript reviews the abrasive water jet machining (AWJM) of titanium alloys, with particular emphasis on Ti-6Al-4V, by systematically analysing the effects of key process parameters, individually and in combination, on machining quality indicators such as surface roughness and material removal rate. The review critically evaluates the existing parameter-optimisation techniques to identify the dominant factors influencing machining performance. Importantly, this review uniquely incorporates a focused and critical discussion on the application of artificial intelligence and machine learning in the AWJM of titanium alloys, an emerging area that has not been comprehensively addressed

https://doi.org/10.3390/designs10010013

in previous reviews. By integrating conventional process-parameter studies with emerging data-driven modelling and optimisation approaches, this review fills a clear gap in the literature and provides practical guidance for researchers and engineers seeking to enhance the machining performance of Ti-6Al-4V.

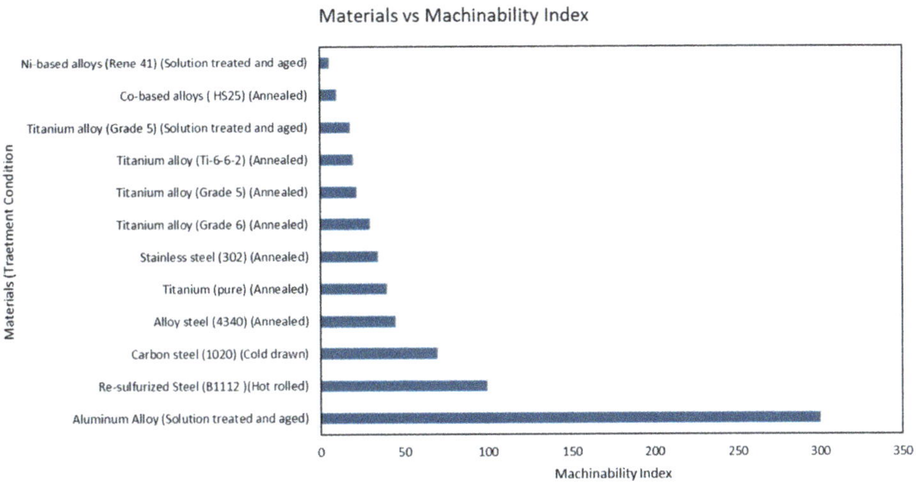

Figure 1. Machinability index of different materials. The figure was drawn based on the information from the literature [14].

2. Abrasive Water Jet Machining (AWJM) of Titanium Alloys

The traditional machining processes of titanium alloys possess several limitations. The heat formation during machining gives rise to changes in microstructure within the machining zone "due to the low thermal conductivity of Ti-6Al-4V and the formation of the built-up edge (BUE) chips that are detrimental to the machined surface quality" [15,16]. Other limitations of the "conventional machining processes, such as milling and drilling, were high tool wear, low material removal rate (MRR)", low dimensional accuracy, and high surface roughness [17–19]. All of these limitations open the door for alternative machining practices for the titanium alloy, commonly known as non-traditional machining, which includes, but is not limited to, "electro-discharge machining (EMD)" [20,21], "electro-chemical machining (ECM)" [22], "ultrasonic machining (UM)", "laser beam machining (LBM)" [23], "water jet machining (WJM)" [24], and "abrasive water jet machining (AWJM)" [25]. In contrast to "conventional machining," no cutting tools were used for material removal in this process. Thus, this process does not induce shear deformation, the material deformation mechanism, during conventional machining [26,27]. The material removal mechanisms are "different in non-conventional machining and conventional machining. In the former, residual stress and chatter while machining can also be overcome" [28]. Out of these different "non-traditional machining" practices, AWJM of Ti-alloys is one of the promising processes that is foreseen to provide quality and efficient machining of complex geometries of titanium alloys [29].

AWJM uses a "high-velocity jet of water mixed with abrasive particles to cut materials". This material can be alloys, metals, ceramics, composites, and even "thick materials that are difficult to machine using conventional methods" [30,31]. AWJM relies on the fundamental physics of fluid mechanics and erosion. The process involves accelerating water to supersonic speeds using a high-pressure pump, typically in the range of 200–600 MPa. The water jet is then mixed with abrasive particles such as garnet or aluminium oxide,

https://doi.org/10.3390/designs10010013

which are introduced into the jet stream through a mixing chamber. When the high-speed mixture impacts the workpiece, the "kinetic energy of the abrasive particles" converts into mechanical energy, causing "material removal through a combination of micro-cutting and erosion". A schematic of the simplified process of this method was depicted in Figure 2.

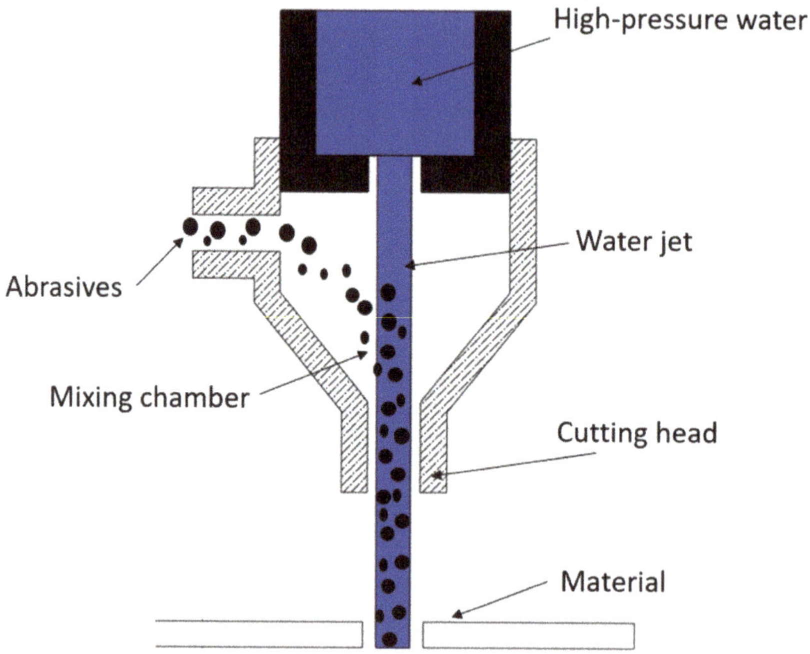

Figure 2. A simplified schematic of the process.

The cutting process in AWJM involves several physical phenomena that take place simultaneously: namely, (i) fluid dynamics (Bernoulli's principle), (ii) abrasive particle velocity, and (iii) different material removal mechanisms that are relevant to the "workpiece" material [32]. As the high-pressure jet is forced through a small-diameter nozzle, commonly known as an orifice (typically 0.1–0.4 mm [33]), the water forms a jet and exits at a significantly high velocity (up to 900 m/s [34]), accordingly to Bernoulli's equation in Equation (1):

$$P + \frac{1}{2}\rho v^2 + \rho g h = constant \quad (1)$$

where P = pressure, ρ = fluid density, v = velocity, g = acceleration due to gravity, and h = height.

After forming the desired water jet, abrasives are introduced into the jet stream via a venturi effect, where the high-speed water creates a low-pressure region, drawing in the abrasive particles [35]. The abrasives mixed with the water jet gain substantial kinetic energy and are focused towards the "workpiece" in a precise way. Upon striking the "workpiece" material, the controlled erosion of the material took place according to the inherent nature of the material [36]. The mechanism can be in the form of brittle and quasi-brittle fracture for "brittle materials" [37] such as ceramics and glass. The impact of abrasive particles on the material surface induces localised fracture when the compressive stress within the plastic deformation zone exceeds the material's tensile strength. The

 https://doi.org/10.3390/designs10010013

resulting radial crack propagation, oriented perpendicular to the surface due to the high stress concentration, subsequently triggers the formation of lateral cracks that are parallel to the machined surface, which are primarily responsible for material removal [38,39]. In the case of ductile materials (e.g., metals, plastics), plastic deformation and cutting prevailed, where repeated impact from abrasives causes plastic deformation, micro-cutting, and progressive layer-by-layer removal of the workpiece material [25]. Figure 3a,b show the material removal in brittle material and ductile material, respectively.

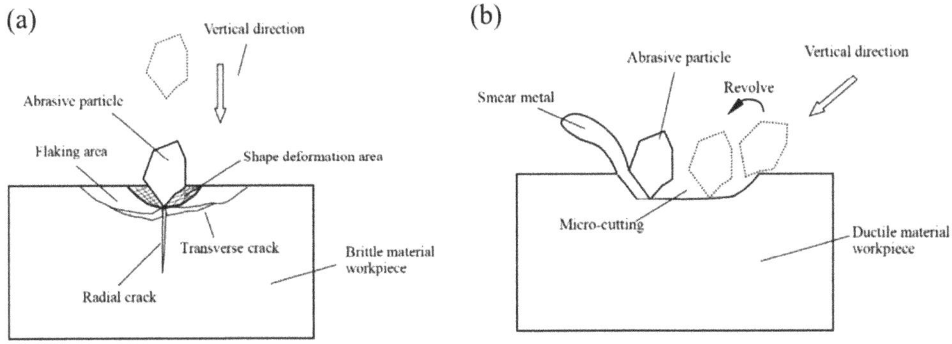

Figure 3. Material removal mechanism in (**a**) brittle material and (**b**) ductile material [39].

The scenario is different for polymer matrix composite materials, where abrasives cut through fibres and matrix phases differently, leading to controlled shearing [40]. The erosion kinetic energy (U_K) and effective erosion kinetic energy (U_M) are as in Equations (2) and (3), below [41].

$$U_M = \frac{U_K}{1+\sqrt{\frac{H}{H_a}}} \tag{2}$$

where U_K—erosion kinetic energy and Ha—hardness of abrasive particle.

$$U_K = 1/2mv_a^2 \tag{3}$$

where v_a—velocity of particles.

At the bottom, there is a catcher that dissipates the remaining energy of the jet after it passes through the "workpiece". The whole process is computerised and controlled automatically by a mechanical/robotic arm in a closed chamber. Both the water and abrasive were collected at the end of each run and recirculated in the system. As it is a closed-loop process, the recovery of both the water and abrasive is high. The abrasive could be used for several runs until it becomes fragmented and loses its erosion capability.

As an AWJM is carried out, the deflection of the jet through the material being cut was observed [42]. Upon entering the material, the water jet will deflect in the opposite direction to the nozzle motion direction. The jet exiting the material "lags behind the point at the top of the material where the jet enters". The horizontal distance from the exit point to the entry point is called the trail back (σ). The jet moves in the direction of the arrowhead in Figure 3. It was noticed that "the jet–material interface is a curved surface". These phenomena are shown schematically in Figure 4.

The angle the water jet deflects at, relative to the perpendicular axis at the water jet entrance, is the declination angle, as shown in Figure 5. The trail back is formed due to the declination angle and the curvature of the path between the "inlet surface and the outlet surface of the cut". The recommended declination angle for material with 30 mm thickness

https://doi.org/10.3390/designs10010013

is 45°, material with 60 mm thickness is 22.5°, and material with 120 mm thickness is 15°. For thickness above 120 mm, the declination angle should be below 10° [43].

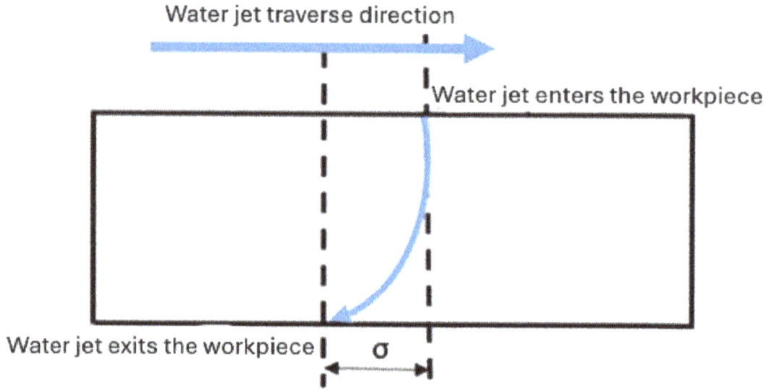

Figure 4. Schematic of the trail back formation during AWJM.

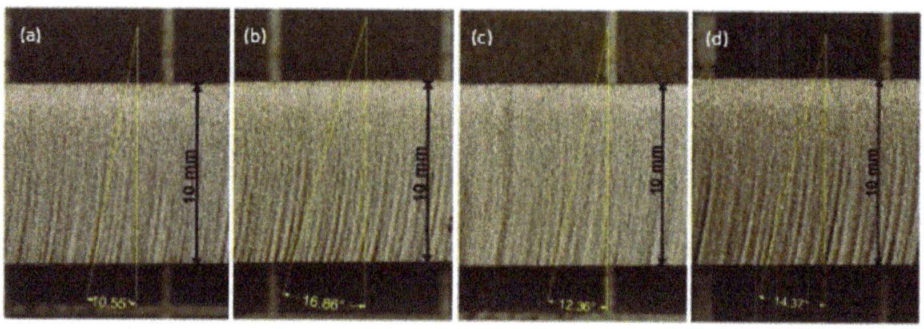

Figure 5. Average angles from left to right: (**a**) high strength steel, (**b**) tool steel, (**c**) stainless steel and (**d**) Hardox 500 [43].

3. AWJM Process Parameters

The AWJM has several parameters that govern the quality of machined components. Figure 6 illustrates the various input parameters and the output parameters that directly influence the quality and productivity of the process.

3.1. Hydrodynamic Parameters

The use of WJM to "machine gamma titanium aluminide (γ-TiAl)" reported that the water jet pressure and jet paths affected the generated surface [44]. It was observed that a lower water pressure does not form continuous erosion traces on the workpiece surface. The erosion traces turn out to be more continuous with the rise in water pressure. With a further increase in jet pressure, erosion rises along the depth and develops continuously along the jet path [44]. Higher water pressure increased the jet velocity and kinetic energy of the water jet, enhancing the cutting efficiency [45].

During the AWJM of Ti-6Al-4V, three different microstructural regions form in the "workpiece" material: (a) the "initial damage region" (IDR), (b) the "smooth cutting region" (SCR), and (c) the "rough cutting region" (RCR) [46]. The IDR is the "deformation on the surface of the material due to the successive impact of particles from the water jet", the SCR is the surface area with reduced "roughness due to the water jet" retaining sufficient kinetic energy, and RCR is the surface area with the most roughness "due to the water jet delay at the material's exit" [47]. A surge in jet pressure increased the microhardness in all three

 https://doi.org/10.3390/designs10010013

microstructural regions due to "plastic deformation and strain hardening". The increase in hardness of the cut surfaces was also attributed to the embedment of abrasive particles, which will be discussed in a later section [48]. The increase in water pressure was also found to result in the formation of non-recrystallised grains and deformed grains and increased the dislocation density [49]. The increase in the dislocation density induced the "embrittlement in order to reduce ductility". During hole-machining, the "depth of cut" and the diameter of the hole are directly related to the jet pressure, and the surface roughness is inversely related to the jet pressure [37,50]. An upsurge in jet pressure increased the abrasives' kinetic energy, leading to a higher depth of cut. In a pocket milling study for Ti-6Al-4V, at lower jet pressures, the "milled surface was found to be uniform with good surface finish for all traverse speeds". As the pressure increased, the surface became uneven and rough due to excessive material removal. The surface was found to "contain grooves and ridges at high jet pressure" [51]. To summarise, an increase in water jet pressure results in a greater depth of cut and higher material removal rates, due to the increased kinetic energy of the water jet. Increasing the water jet pressure up to an optimum level has a positive effect on surface roughness, due to a more uniform material removal. However, excessively high water jet pressure leads to increased surface roughness and waviness due to over-erosion; therefore, it must be carefully balanced in conjunction with other process parameters, as discussed later.

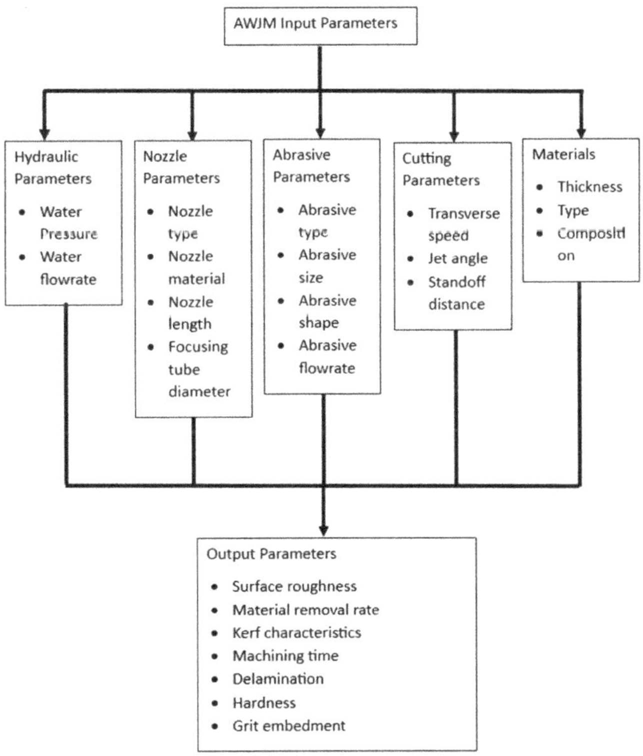

Figure 6. AWJM input and output parameters.

3.2. Nozzle Parameter

The crucial components of the AWJM system are the orifice and nozzle, which form water jets, as shown in Figure 7. The nozzle diameter and shape influence jet focus and precision, with

 https://doi.org/10.3390/designs10010013

smaller nozzles enabling more detailed cuts through energy conversion from pressure energy to kinetic energy that is transmitted to the water jet and abrasives [52]. The recommended orifice range is 0.13 to 0.76 mm [42]. The nozzle has three important parameters: "convergence tube angle (α), abrasive incidence angle (θ) and abrasive feed length." The smaller convergence angle provides more axial acceleration to the abrasives, hence increasing the kinetic energy of the water jet. Abrasive particles achieve high velocity in the mixing chamber, under a "minimum abrasive incidence angle". The position of the abrasive feed tube should not be close to the lower section nozzle for the abrasives to gain a higher initial velocity due to the Venturi effect [53]. The nozzle is a consumable, which is made of hard materials like "sapphire, ruby, natural diamond, or synthetic mono- or polycrystalline diamond" to minimise/resist wear by the high-velocity water jet. The edge of the nozzle must be sharp, with a minimal fillet radius, to form high-coherency water jets. The formation of different jet coherency was shown in Figure 8 at 350 MPa WJP [42]. The ideal jet will be the laminar flow where the jet strikes the workpiece in the focused region with maximum kinetic energy.

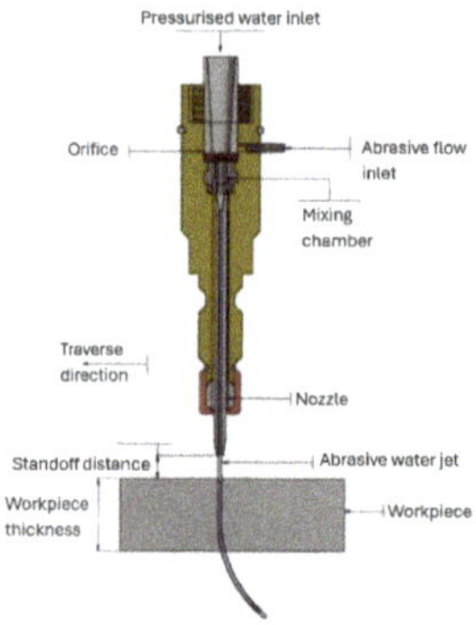

Figure 7. AWJM nozzle components [54].

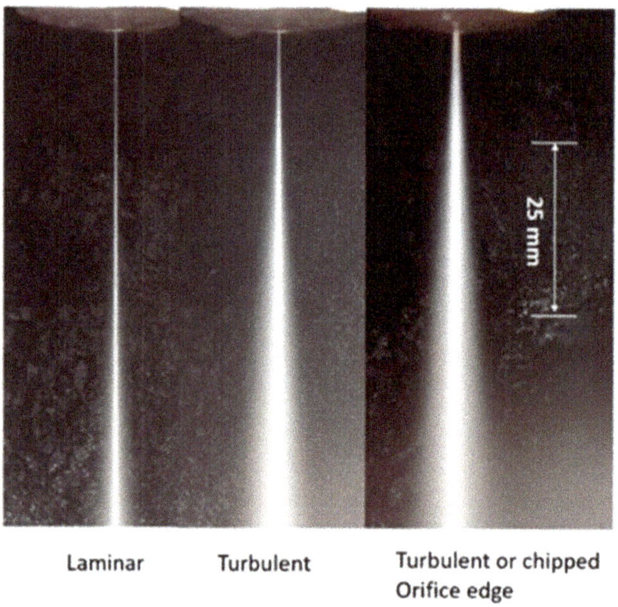

Figure 8. Different water jet flow at 350 MPa pressure [42].

https://doi.org/10.3390/designs10010013

Computational fluid dynamics (CFD) studies on erosion of the internal components of abrasive water jet cutting heads have shown that abrasive particles entrained in a high-speed water jet mix with the turbulent flow within the mixing chamber. The maximum erosion is consistently reported at the junction between the mixing chamber and the focusing nozzle. These studies further confirm that the erosion rate increases with the increasing abrasive mass flow rate and particle size, owing to the higher particle mass and kinetic energy. In addition, the flow pressure within the nozzle is non-uniform along the radial direction: the pressure is lowest and velocity is highest at the nozzle centre, whereas the pressure is highest and the velocity approaches zero near the nozzle wall. An axial pressure drop is also observed along the nozzle length, with higher pressure at the inlet and lower pressure at the outlet. Consequently, the water jet velocity increases along the nozzle axis and reaches a maximum at the nozzle centre [55,56].

3.3. Abrasive Characteristics

The preferred abrasives in AWJM are "garnet, aluminium oxide, glass beads, olivine sand and steel grit". Garnet is commonly used due to its availability, non-toxicity, and mechanical and physical properties. "The most important properties of abrasives are hardness, specific gravity, size, shape, and frangibility". Among them, the "shape of the abrasive particles" plays a dominant role, as this actually acts as a 'cutting tool' during the AWJM process. "The sphericity and roundness, the most commonly known indices, of the abrasive particles" were shown in Figure 9. The sphericity refers to the "extent to which a particle approaches a spherical shape". The roundness relates to the corners and the edge sharpness of an abrasive particle. In the application in AWJM, abrasives with a "high sphericity and low roundness index" are preferred. Glass beads, garnet, brown and white aluminium oxide, steel, olivine, silica, silicon carbide, and zirconia are generally used as abrasives for machining titanium alloys [57,58].

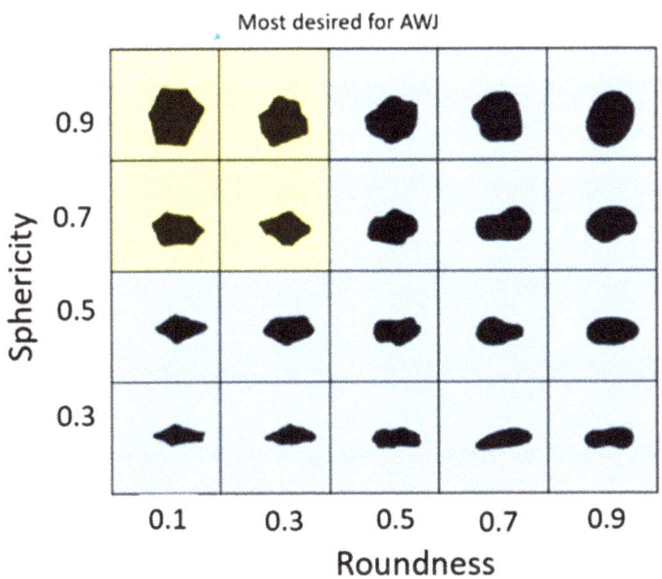

Figure 9. The preferred abrasive shape for AWJM (highlighted in yellow) [42].

 https://doi.org/10.3390/designs10010013

A comparison study between SiC and garnet abrasives reported that SiC produced a greater depth and material removal rate (MRR), with minimised kerf taper angles compared to garnet under similar machining conditions, as SiC is harder than garnet. However, the multi-objective optimisation study of this work concluded that even though SiC is the abrasive of choice for an individual output—for example, MRR—to achieve optimisation for more objectives that might contradict each other (for example, higher depth and lower kerf width), garnet was found to be the preferred abrasive [59]. In spite of the widespread choices, 86% of AWJM processes used a garnet abrasive [60].

Garnet is relatively hard, easily available, and cheap. In addition, this abrasive has sharp edges, better flow characteristics, and "better performance" compared to "silicon carbide, silica, aluminium oxide, steel grit, steel shot, copper slag and glass beads" [61]. The shape factor of abrasives, type of abrasives, and hardness of abrasives are the variables, apart from jet pressure, jet impingement angle, and traverse speed, that affect the AWJM of titanium alloy [57,62,63]. The use of eco-friendly soft walnut as an abrasive was found, under optimum conditions, to be able to be used with harder abrasives for cleaning and polishing operations, leading to higher productivity and surface integrity [64]. The characteristics of the common abrasive used in the AWJM process was tabulated in Table 2.

Table 2. Characteristics of the common abrasive [42].

Abrasive	Specific Gravity	Hardness		Rough Relative Cost	Roundness	Sphericity	Frangibility Level (*)
		Knoop	Moh				
Garnet	3.4–4.3	1350	7.5	1	0.45	0.78	medium
Aluminium oxide	3.95–4.0	2100	9	4–6	0.35	0.78	medium
Silicon carbide	3.2	2500	9.2	3–4	0.31	0.75	medium
Chilled iron	7	520	6	4–5	0.5	0.8	medium
Steel grit	7	500	5	5–7	0.52	0.82	low
Steel shot	7	460	5	4–6	0.89	0.93	low
Copper slag	2.8–3.8	1050	7	0.5	0.5	0.78	high
Silica sand	2.2–2.65	700	7	0.5	0.57	0.79	high
Olivine	3.2–4.5	1100	6.5	0.75–1	0.6	0.82	high
Staurolite	3.7–3.8	1275	7.5	0.6–0.7	0.46	0.79	medium
Glass beads	2.5	700	5.5	1.5–2	0.95	0.95	high
Tungsten carbide	14.92	1870		7–10	0.47	0.77	low

(*) Author's estimate depends on grit size and target material.

The MRR reduced with the rise in the "shape factor" for all traverse speeds (0.03 m/s, 0.083 m/s and 0.166 m/s) [31]. However, the MRR and particle hardness are directly proportional to each other. A sharp rise in MMR was noted, as the particle hardness rises during milling until 100 HV. MRR also increased with an increase in the abrasive size, as high mass abrasives have higher kinetic energy when they impact the material surface [65]. The optimum concentration of abrasives in water was found to be 20% for abrasives such as GMA80, olivine, and glass in the AWJM of Ti-6Al-4V [66]. Above this concentration, the water was not able to provide the maximum velocity to the "abrasive particles" and

 https://doi.org/10.3390/designs10010013

below this concentration, the smaller number of abrasive particles led to a smaller number of cutting blades, hence achieving a smaller cutting depth.

The surface scratch size depends on the abrasive particle size and other parameters, such as "water jet pressure, traverse speed and standoff distance" were observed to have less effect on scratch size [67]. Small particle sizes are suitable for smaller surface roughness in IDR, and large particle sizes are suitable to reduce the surfaces at the SCR and RCR. An increase in "abrasive flow rate and abrasive mesh size" increased the depth of the cut. A larger abrasive flow rate increased the abrasive impingement of the surface, leading to increased surface erosion. The larger abrasive size increased the unit area of the machining surface. The larger abrasive size was also found to increase the top kerf width [68]. A surge in flow rate is directly related to the lowering of surface roughness, as is the increase in "kinetic energy of the jet" and the higher cutting ability of the abrasives. However, observation showed that a reduction in surface roughness will be at the "jet entrance and the jet exit" surface roughness gradually increases due to a "reduction in the kinetic energy of the jet", which is caused by the collision between abrasive particles [69]. The impact of hard particles, as well as the water jet, induces localised plastic strain at the workpiece surface.

Material is removed when this strain surpasses the "strain-to-failure" value of the workpiece material [70]. If the abrasive's hardness is greater than that of the workpiece, an inflexible indent forms, causing plastic strain on the workpiece surface. However, when the particle hardness is lower than that of the workpiece, the particles deform or fracture [57]. Figure 10a,b showed that lower hardness abrasives—for example, steel shot and glass beads (around 500 HV)—gave a better surface finish at different traverse speeds. During the AWJM of titanium alloy, a 50% lesser roughness was generated by the glass bead abrasive compared to that of the garnet abrasive (Figure 10b) [57].

The effects of abrasive grit size on "surface roughness at different traverse speeds" during the AWJM of Ti 6Al 4V are given in Figure 11a [71]. Finer-grit-size abrasives (200# or 75 μm) produced lesser surface roughness compared to the commonly used 80 mesh (80# or 180 μm). Larger grit sizes have the tendency to remove more materials; hence, they will lead to rougher surface finishes. AWJM also induces waviness on the generated surface due to irregularities in the material removal process. Smaller abrasives generate a less wavy surface compared to that of larger abrasives at the same traverse speed, as presented in Figure 11b. This is because of the fact that smaller abrasives induce reduced damage to the surface during machining and are capable of tracking the streamlines of the water jet more accurately.

3.4. *Cutting Parameters*

3.4.1. Traverse Speed

The traverse speed affects the material removal rate (MRR) of Ti-6Al-4V for various abrasives and grit sizes [62,71]. For four different abrasives—garnet, glass bead, Al_2O_3 white, Al_2O_3 brown, and steel shot—at a lower traverse speed of 0.025 m/s, Al_2O_3 brown, with the highest hardness of 1800 HV, produced the highest MRR and glass bead, with the lowest hardness of 500 HV, produced the lowest MRR. At a higher traverse speed of 0.16 m/s, Al_2O_3 white (1800 HV) produced the highest MRR and steel shot (200 HV) produced the lowest MRR. At a range of traverse speed of 0.001 m/s to 10 m/s, the 80-grit abrasives produced higher MRR compared to 200-grit abrasives [57,71]. The ANOVA analysis showed that the water jet traverse speed influences MRR by 84.32% and surface roughness by 54.32% [72]. The MRR for the Ti-6Al-4V alloy can be represented using Equation (4), below [73]:

$$\mathrm{MRR} = \mathrm{h_t} \times \mathrm{W} \times \mathrm{v_f} \quad (4)$$

 https://doi.org/10.3390/designs10010013

where v_f—traverse speed, W—width of kerf (average of top kerf width and bottom kerf width), and h_t—depth of cut.

A non-linear regression equation for MRR, which included the effect of process parameters on MRR, was given by Equation (5) [74]:

$$MRR = -0.2395 + 0.000589\ T_v + 0.00069\ A_f + 0.0861\ S_d - 0.000196\ (T_v \times S_d) - 0.000104\ (A_f \times S_d) \quad (5)$$

where T_v—traverse speed, A_f—abrasive flow rate, and S_d—standoff distance.

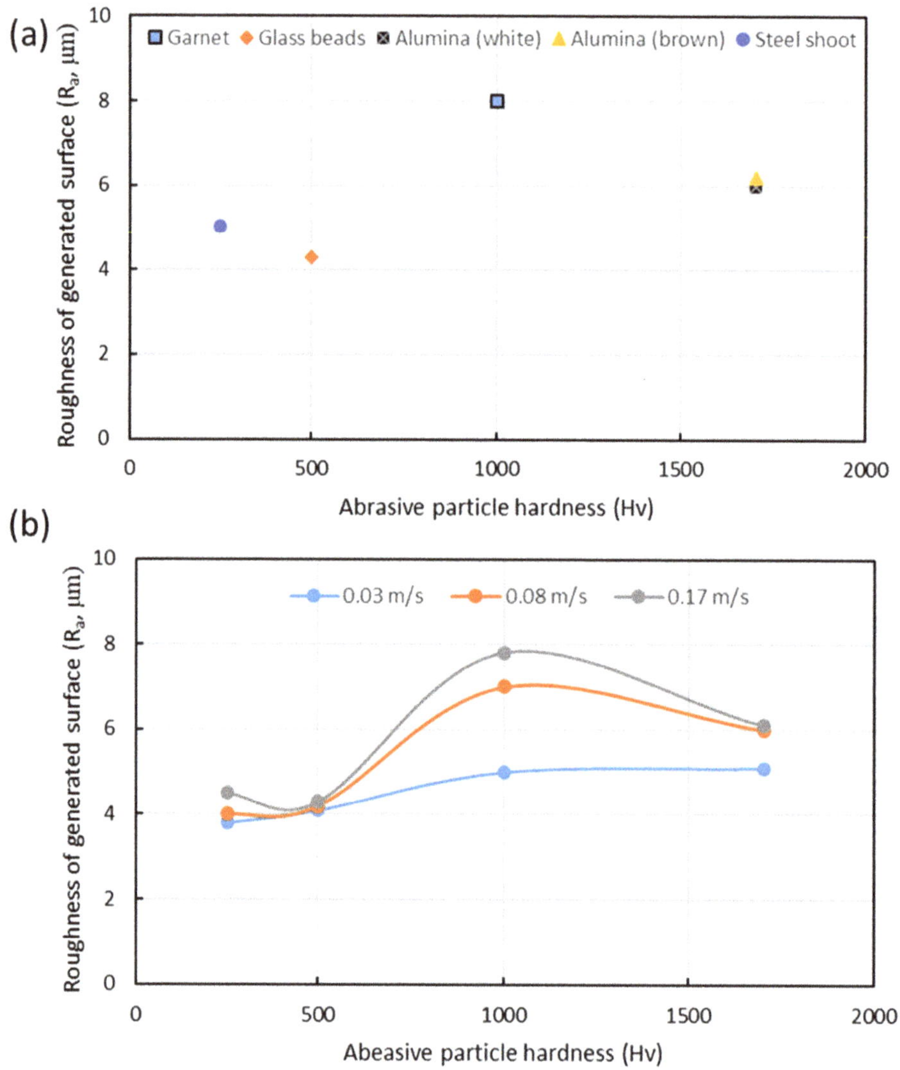

Figure 10. Influence of particle hardness on the roughness of generated surfaces for (**a**) different abrasives and (**b**) at different traverse speeds (m/s). The graphs were drawn based on information in ref. [57].

https://doi.org/10.3390/designs10010013

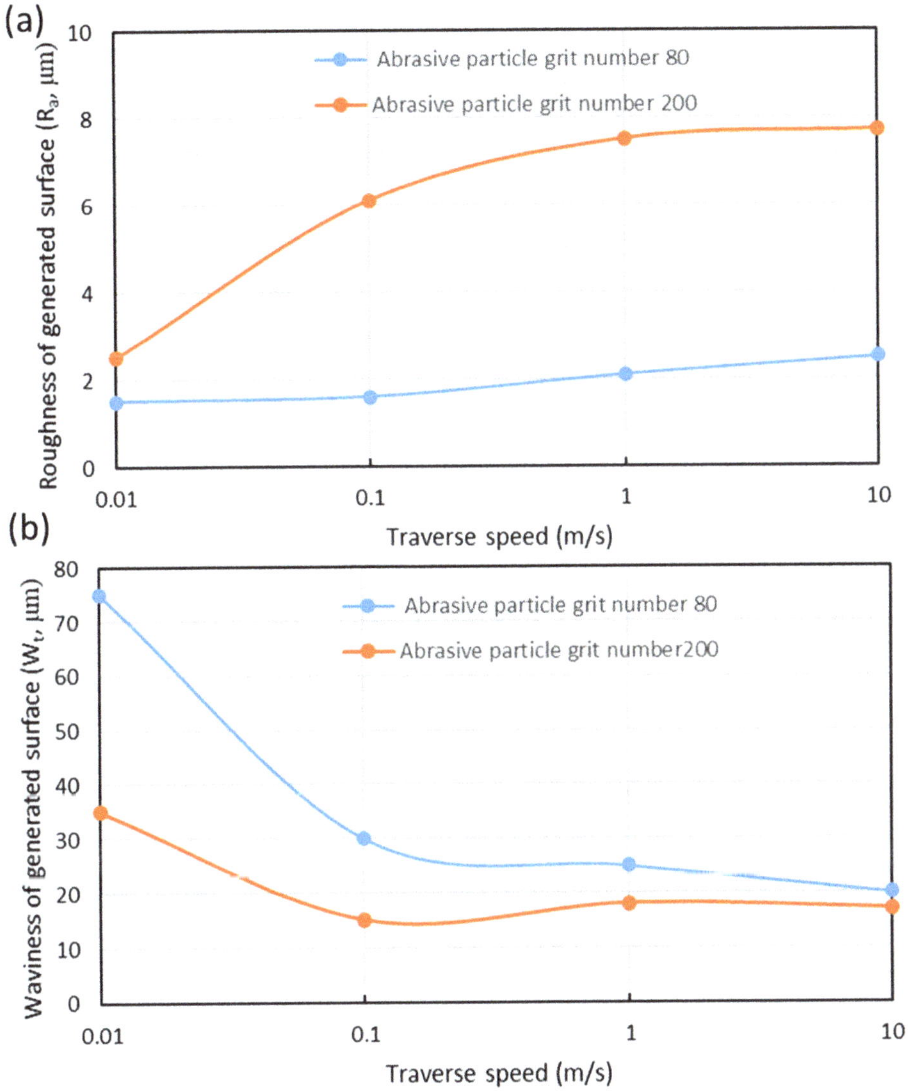

Figure 11. Effect of different abrasive particle grit size on (**a**) surface roughness and (**b**) surface waviness. The graphs were drawn based on information in ref. [71].

The increase in traverse speed increased the MRR but also increased the surface roughness, as abrasive particles do not have sufficient time to remove surface asperities from the surface to produce enhanced surface quality [74]. The increase in traverse speed is inversely proportional to the "depth-of-cut" because of the decrease in jet kinetic energy, due to the reduction in time to carry out a cut at a point [75].

It was found that the IDR decreased slightly, and the SCR decreased significantly with the increase in "traverse speed, which reduced the intensity of particles hitting a target area and the depth of penetration". The texture of the generated surface indicates that material removal in AWJM occurs by the amalgamation of two actions, such as "scooping-induced ductile shear and ploughing actions by the abrasives" [63,76]. According to Wang et al. [77],

 https://doi.org/10.3390/designs10010013

"the material removal mechanisms of AWJM in the IDR and SCR are recognised as cutting and deformation", correspondingly, while RCR is contributed by erosive wear at a large abrasive attack angle [38]. In the erosion process, abrasives push the material on the surface in the direction of focus (against the surface) because of abrasive deceleration, and the surface suffers elastic and plastic deformation as a response to the external force [57]. The magnified pictures of the generated surface shown in Figure 12 indicate that the amount of "plastic deformation" rises from the "top to the bottom of the machined surface". In the IDR of the machined surface, abrasives induce significant damage due to significantly greater kinetic energy [63]. When the abrasives penetrate more into the material, the "kinetic energy of particles" reduces, and part of it is consumed to form the SCR. Therefore, a lower energy abrasive stream generates RCR where the stream deflects from the normal direction to the cutting plane and forms striation on the machined surfaces [63]. The "declination angle of striations" on the machined surfaces is generated during their interaction with the workpiece material, which is larger than the "angle of the relevant faster jet core" [78]. With the increase in traverse speed, the amount of deflection rises, and the striation generation is affected by the decrease in stream energy and the striation angle. The striation formation depends on the dynamics of the water jet, the type of abrasives, and the machine vibration [79].

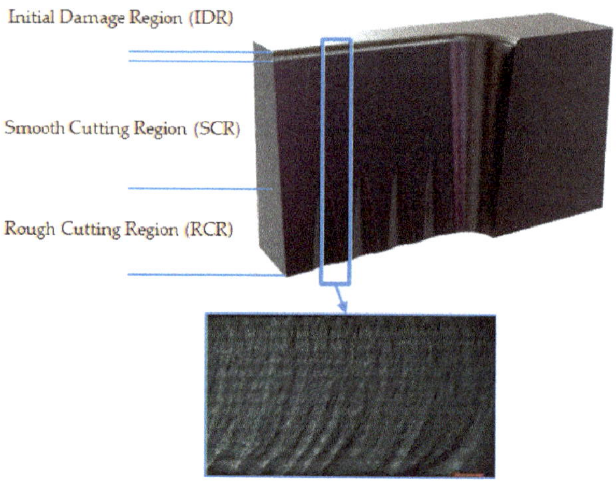

Figure 12. Different surface roughness zones after AWJM: IDR, SCR, and RCR [80].

3.4.2. Stand-Off Distance

As the standoff distance increased, the abrasive-laden water jet diverged, creating a "wider and less focused cutting zone". This increased the MRR [65]. Even though a higher standoff distance increased the MRR, it affected the surface roughness due to an increase in random surfaces consisting of peaks and valleys [81]. In contrast, reduction in "standoff distance decreased surface roughness" due to the targeted abrasive stream, leading to hardening of the workpiece surface [82]. The optimum lower standoff distance has been reported as 2 mm [83]. The abrasive water jet is made of two zones: the inner zone (a), with higher velocity particles, and the outer zone (b), with low velocity particles. The increase in "standoff distance" increases the outer zone compared to the inner zone, leading to divergence in the Gaussian distribution of particles [84]. The increase in standoff distance is "inversely proportional to the depth of cut" for the same reason as the "increase in traverse speed"; reduction in jet kinetic energy causes insufficient energy to "increase the depth of cut" [75]. The same reason applied to the "increase in surface roughness with an

 https://doi.org/10.3390/designs10010013

increase in stand-off distance; the lower density abrasive particles due to the jet expansion, generate more random peaks and valleys on the surface" [74,85]. The "kerf taper increased as the standoff distance" increased from 2 to 10 mm at different traverse speeds. However, with the "increase in standoff distance at different increasing traverse speeds", the entrance and exit widths decreased but the kerf taper angle increased [86].

3.4.3. Jet Angle

The "crater depth" for the Ti alloy, due to the impact of abrasive particles, has been reported to increase with an increase in the jet angle (30° and 60°), the number of particles, and increased particle velocities of 180 m/s, 200 m/s, and 220 m/s [70]. The depth of cut is higher for the 60° jet angle compared to the 30° angle, as at a "higher velocity and impact angle", the particles gain higher energy and momentum; therefore, the "deformation on the workpiece increases". In the study of jet angles (50° to 90°) and two pitch values of 0.7 mm and 1.1 mm, the depth of cut increased with an increase in the jet inclination angle, but for the same jet angle, the pitch of 0.7 mm produced deeper cuts compared to the pitch of 1.1 mm [87].

4. Machining Mechanism and Parameter Optimisation

The multi-pass strategy for the AWJM of the Ti-6Al-4V alloy was reported to assist in increasing the "depth of cut", with the water jet pressure and traverse speed playing major roles in achieving the desired depth. Even though the strategy did not have a major impact on the top kerf width, as the kerf width showed a slight "increase with a higher number of passes", the method effectively reduced the kerf angle after multiple passes. Similarly, the MMR and cutting efficiency also decreased with increases in passes, hence showing that the single-pass method is preferred for higher MRR and efficient cutting [88].

The roughness of the surface generated from the AWJM depends on the traverse speed [89], properties of abrasives, jet pressure [57,71], standoff distance [89], and abrasive flow rate, which are the most substantial variables to determine the "quality of the machined surfaces". A greater rate of abrasive flow with longer standoff distance produces rougher surfaces because of higher and more random energy dispersal [90]. "The surface roughness increased with the increase in jet traverse speed" [91,92]. With an increase in traverse speed from 0.03 to 0.166 m/s, surface roughness increases by 20% for most of the abrasive particles, although it increased by 40% for garnet abrasives [57]. This might have taken place because of fewer machining action overlaps and fewer abrasives to impact the surface as the traverse speeds rose [93]. Hascalik et al. [63] reported that the "depth of cut" increases and surface quality worsens because of the reduced "kinetic energy of the jet stream". The jet impact angle degrades in RCR, and the disfigurement wear mechanism becomes more effective than at the highest position of the machining zone, where the cutting wear mechanism is predominant. The average surface roughness of 2.30–3.57 μm at 1 mm depth can be achieved at 60–250 mm/min traverse speeds, respectively [63]. The Ra values between the IDR, SCR, and RCR, with the RCR having the highest Ra value, showed that the deflected water jet creates a secondary material removal with a reduced momentum. This secondary material removal predominantly acted at the IDR and SCR [94].

4.1. Experimental Design for Parameter Optimisation

A number of studies have used the "Taguchi method for process optimisation" of AWJM to machine titanium alloys (Ti-6Al-4V). "The increase in pressure and standoff distance affected the surface roughness and hardness, and the influence of abrasive flow rate and traverse speed is low for both output parameters" [95]. The increase in traverse speed and the abrasive flow rate with sharp edges significantly increased the material

https://doi.org/10.3390/designs10010013

flow rate. However, the increase in traverse speed increased surface roughness, while the increase in the sharp-edged abrasive flow rate decreased the surface roughness. The roughness Ra during the AWJM of Ti6Al4V is represented by Equation (6) [69]:

$$Ra = 61.38 + 0.00624667\ A - 0.1093\ B - 0.0105522\ C - 31.67667\ D - 1.4\ E \\ -0.06\ A^2 + 0.00001330\ B^2 + 4.12222\ E - 0.06\ C^2 + 6.62\ D^2 \quad (6)$$

where A stands for water jet pressure, B stands for abrasive mass flow rate, C stands for traverse speed, and D stands for standoff distance. Another surface roughness control model is given in Equation (7) [96]:

$$Ra = 4.68 + 0.24\ B + 0.22\ AB + 0.18\ AC - 0.14\ BC - 0.24\ A^2B + 0.16\ A^2C + 0.088\ A^3 \quad (7)$$

where A stands for travel speed, B stands for water pressure, and C stands for standoff distance. The study that report Equation (7) confirmed that the cube of travel speed (A^3) and the interaction between travel speed and pressure (AB) are the most significant terms for the surface roughness. Both Equations (6) and (7) showed the combined effect of the input parameters.

ANOVA with the smoothed particle hydrodynamics (SPH) approach to model the AWJM [97] reported that the residual stress on the machined component and surface roughness can be optimised with the combination of parameters such as water jet pressure, abrasive flow rate, traverse speed, orifice diameter, standoff distance, abrasive type, and mesh. The choice of abrasive type and size is also crucial, as harder abrasives such as garnet improve the cutting capability, while finer particles result in smoother surface finishes [98]. The standoff distance, or the gap between the nozzle and workpiece, affects the jet dispersion and cutting accuracy [99], while the cutting speed determines the material removal rates, with higher speeds increasing productivity but reducing depth penetration [100]. A Taguchi and ANOVA parameter optimisation study also reported that among the parameters of water pressure, standoff distance, and nozzle diameter, the nozzle diameter significantly affects the MRR while the quality of the hole is mostly influenced by the standoff distance [101]. The standoff distance was also reported to be the highest influence parameter between the water pressure, abrasive flow rate, feed rate, and standoff distance [102]. This Taguchi–DEAR process optimisation study identified that the standoff distance has the highest contribution to the material removal rate, surface roughness, and multi-response performance index in titanium alloy machining, as the standoff distance is directly related to the impact energy that produces the surface craters on machined surfaces.

A Grey Rational Analysis combined with Taguchi method optimisation study [103] on the AWJM of Ti6Al4V reported that the depth of cut is most controlled by the traverse speed followed by the abrasive mass flow rate. Regarding the interaction between traverse speed and mass flow rate, a lower traverse speed and higher mass flow rate will produce the highest depth of cut. This study also reported that the surface roughness was highly impacted by the water pressure, followed by the traverse speed. Surface roughness can be improved by higher water pressure and lower traverse speed. Another Grey Rational Analysis combined with Taguchi method optimisation study showed that, in AWJM slot machining of Ti6Al4V, the water jet pressure has the highest significance of 55.4%, followed by the traverse speed at 28.57%, and the abrasive mass flow rate at 16.06% [104]. This study used glass beads as the abrasive particle, due to their higher recyclability and lower ability to create surface damage compared to garnet abrasives. The study that used SiC in garnet as abrasives reported that the Grey Relational Approach–Taguchi Method study showed that the SiC content, SiC size, and abrasive flow rate were the optimal process parameters [105]. The addition of SiC in garnet improved the material removal rate as the hexagonal crystal

https://doi.org/10.3390/designs10010013

of SiC increased its hardness, hence removing materials at high velocity via brittle fracture. However, the addition of more SiC particles increased the surface roughness as the particles collided with neighbouring particles, creating large criss-crosses on the workpiece surfaces. The Taguchi optimisation study on titanium 6242 reported that traverse speed contributes about 98.2% to the MRR and 33.5% to the surface roughness compared to the water pressure and standoff distance. Both the traverse speed and standoff distance have almost similar contributions to the kerf angle, with 31% and 37%, respectively [106].

A "multi-objective optimisation algorithm (MO-Jaya algorithm) was used to optimise the surface roughness, milling depth and material erosion rate in AWJM of titanium Ti-6Al-4V". Milling depth is negatively correlated with traverse speed, because an increase in speed reduces the erosion time, leading to a decrease in depth. The effect of the feed rate on the depth is the same as in traverse speed; an increase in the feed rate shrinks the jet overlap area. Deeper milling depth can be produced with a greater abrasive flow rate. The mass flow rate has the most significant effect on the MRR, followed by jet pressure. As for surface roughness, the traverse speed is more sensitive to changes in surface roughness compared to other parameters. Surface roughness decreases linearly with an increase in both the traverse speed and standoff distance [107]. The VIKOR method, a multi-criteria decision-making method used in the AWJM of titanium alloy, concludes that the pressure is the most important parameter, and an optimum traverse speed will produce minimum surface roughness [108]. Another multi-objective optimisation study that used a grey relational study reported that the radial overcut of a hole increased with the standoff distance. At lower pressure and lower standoff distance, the radial overcut is less, as the abrasive particles impinge on the workpiece with rebounding. The taper angle also increased with an increase in standoff distance, due to the flaring of the water jet. Finer particles produced more taper than the coarser particles, as coarser particles retained their energy at a higher depth while the finer particles lose energy with an increase in depth, leading to an increase in taper. The GRA optimisation showed that the hole properties, such as radial overcut and taper angle, are affected by the standoff distance by 67.3% and the abrasive size by 15.3% [109]. Table 3 provides a summary of parameter optimisation study outcomes for the AWJM of Ti-6Al-4V.

Table 3. Parameter optimisation study for Titanium alloys.

Reference	Technique	Significant Parameter and Percentage
[72]	Taguchi method	Traverse speed—54.53% significant for surface roughness and 84.32% significant for MRR.
[50]	Taguchi method	Traverse speed—90% significant for depth of cut.
[90]	Taguchi method	"Abrasive flow rate and standoff distance most significant for surface roughness".
[101]	Taguchi method	Nozzle diameter is most significant for MRR. Standoff distance is the most significant for kerf accuracy.
[96]	Response surface method	"Traverse speed and water pressure" are significant (as a single factor) for surface roughness.
[73]	Response surface method	"Water jet pressure and traverse speed are significant for kerf geometry and surface roughness. Water jet pressure, traverse speed and abrasive flow rate are significant for the material removal rate".

 https://doi.org/10.3390/designs10010013

Table 3. *Cont.*

Reference	Technique	Significant Parameter and Percentage
[102]	Taguchi–DEAR method	Standoff distance the most significant for MRR and surface roughness.
[66]	Response surface method	"Traverse speed is the most significant for the depth of cut".
[65]	Multivariate regression model	"Water jet pressure and mass flow rate are the most significant for the depth of cut".
[103]	Grey relational analysis + Taguchi method	"Traverse speed is the most significant for depth of cut followed by abrasive flow rate. Water pressure is the most significant for surface roughness, followed by traverse speed".
[110]	Box–Behnken design method + response surface method	"Traverse speed is the most significant for milling depth and surface roughness followed by water pressure. The combined effect of abrasive flow rate and standoff distance is significant for surface roughness".
[95]	Taguchi method	Water jet pressure is the most significant for machining duration, surface roughness, and surface hardness.
[68]	Taguchi method	Abrasive flow rate is the most significant for depth of cut, with 61.53%, and material removal rate with 61.82%.
[48]	Taguchi method	Water pressure provides the highest contribution to machining time (61.47%) and surface roughness (57.17%), while the abrasive flow rate has the highest significance for surface hardness (73.7%).
[111]	Grey entropy weigh method	"Standoff distance is the most significant followed by abrasive flow rate for material flow rate, surface roughness and kerf angle".
[112]	Taguchi method	"Water pressure is the most significant (92%) for surface roughness".
[113]	Multi-criteria decision-making (MCDM)	"Water jet pressure is the most significant for depth of cut, surface roughness and material removal rate followed by traverse speed".

4.2. Application of Artificial Intelligence (AI) in the AWJM of Titanium Alloys

The use of "artificial intelligence and machine learning" is a current trend for identifying an optimal parameter combination in AWJM of composites and metal alloys. Studies have shown that a backpropagation artificial neural network (ANN) is a suitable model for faster computation and increased accuracy [114–117]. Figure 13 shows an ANN model for AWJM.

The backpropagation ANN with one hidden layer of 5-neurons was used to predict the surface roughness during the AWJM of a titanium alloy. The ANOVA study showed that the water jet pressure was the most important parameter, followed by the traverse speed. The standoff distance was found to be insignificant because of its small contribution to the surface roughness. The ANN machine learning (ML) model was reported to predict the surface roughness with minimal error compared to the support vector machine (SVM) ML and the regression analysis statistical method [118]. Two backpropagation ANN models

https://doi.org/10.3390/designs10010013

with one hidden layer, one with water pressure and traverse speed as inputs, and the other with mass flow rate and standoff distance as inputs, were developed. The output for both models was the depth of cut [75]. Both models were able to predict the depth of cut with an accuracy of 90%. A hybrid machine learning model was developed using ANN for kerf quality optimisation of AWJM of hybrid stacks of Ti-6AL-4V alloy/CFRP/Al7075 [119]. The ANN has three inputs (jet pressure, traverse speed. and standoff distance), one hidden layer with 17 neurons, and an output layer with nine targets (kerf angle, surface roughness, and material removal rate for all three materials). The ANN model was able to effectively predict the non-linear relationship between the input parameters and the output parameters. The study's objective was to reduce the kerf angle and surface roughness while increasing the material removal rate on the Ti-6Al-4V/CFRP/Al7075 stack, which showed that pressure made a significant contribution of 53.3%, followed by standoff distance (20.49%) and traverse speed (13.52%).

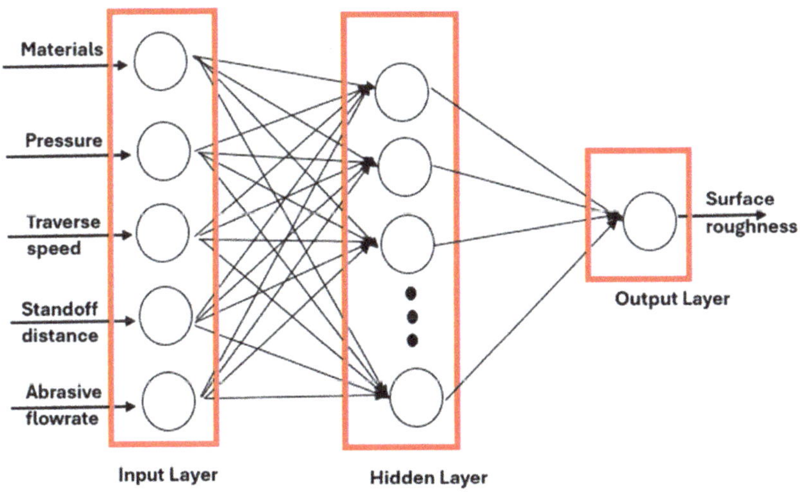

Figure 13. ANN model for AWJM.

A 4-3-1 (four neurons as inputs, three neurons in a single hidden layer, and one output neuron) ANN model with an "artificial neural fuzzy logic algorithm" was created to study the effect of "water jet pressure, standoff distance, abrasive flow rate and traverse speed on the depth of cut" of the titanium metal matrix composite (TiMMC) [120]. The study concluded that a neuro-fuzzy method effectively eliminated the uncertainties in the empirical data and required minimal data for the parameter optimisation study. The method successfully predicted travel speed as the most significant parameter, with a 60.98% contribution, followed by water pressure. "The adaptive neuro-fuzzy inference system, or ANFIS", which includes the learning ability of ANN and the fuzzy logic's ability to deal with uncertainties and vagueness of data, was able to provide closer multi-parameter prediction in the AWJM of Ti-6Al-4V [121]. The ANOVA study in the work also reported that the water pressure was the most significant parameter, followed by the traverse speed. Similarly, Jitendra and Basha [122] used an adaptive neuro-fuzzy system to optimise the parameters of AWJM for cutting Ti-6Al-4V and reported that the ANFIS-reptile search algorithm integration (ANFIS-RSA) was able to improve the prediction accuracy for the surface roughness much better compared to the "ANFIS, ANFIS-PSO and ANFIS-BAS models". The use of a backpropagation ANN to predict the six parameters for a depth of

https://doi.org/10.3390/designs10010013

cut on Ti-6Al-4V, machined using AWJM, was studied in [123]. A 1-10-5 architecture was used, with 10 neurons at the hidden layer and five outputs: water jet pressure, traverse speed, nozzle diameter, abrasive flow rate, and orifice size. The ability of the neural network to accurately predict the process parameters for a given depth of cut was validated using the experiments generated using the Taguchi method. The study showed the ANN algorithm faced difficulty in predicting parameters for a depth of cut of 3 mm to 5 mm, but the prediction improved for higher depth of cuts. A hybrid "Taguchi–Gray Relational Analysis (GRA)" combined with ANN was used to process optimisation for the AWJM of Ti-6Al-V6, where the ANN served as a digital twin to accurately predict machining outcomes based on an experimental dataset generated using Taguchi and GRA [124]. The ANN significantly predicted the new parameter combination, which developed a higher Grey Relational Grade (GRG) than the parameters suggested by the Taguchi Method. This study demonstrated ANN's generalisation and predictive efficiency, reducing the need for a large number of experiments. This study further revealed that SiC addition to garnet abrasives and abrasive flow rate is a significant factor in improving MRR and reducing surface roughness.

5. Discussions and Challenges

Based on the literature review presented above, the AWJM of titanium alloy, such as the Ti-6Al-4V, is governed by different parameters that have a close inter-relationship that affect the selection of other parameters. The optimisation of parameters in the AWJM of titanium alloys involves a crucial balance of key parameters, such as water jet pressure, traverse speed, standoff distance, and abrasive flow rate. The water jet pressure and nozzle traverse speed are the two parameters that have been widely reported in Sections 3.1 and 3.2 as parameters with significant contributions to the AWJM of titanium alloy. These parameters played an important role in maximising the MRR, depth of cut, kerf angle accuracy, and surface hardness and minimising surface roughness in Ti-6Al-4V. The standoff distance and abrasive flow rate have also been reported to play a major role in achieving quality machined surfaces. Figure 14 gives the most significant parameters, based on the 35 process parameter optimisation studies reviewed in this paper. The majority of the reviewed works have reported water pressure, followed closely by traverse speed, as the most significant parameter governing the AWJM of titanium alloys.

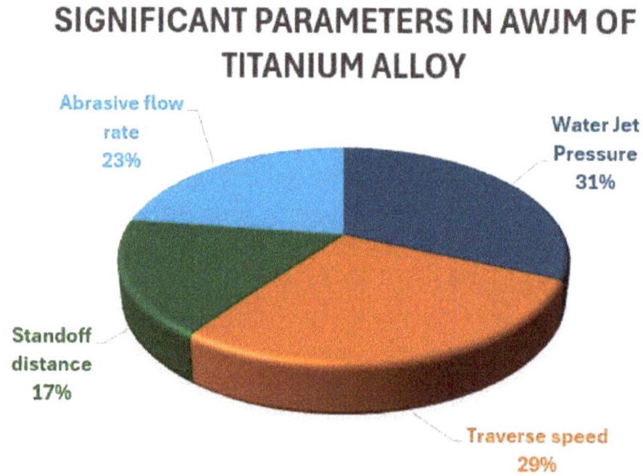

Figure 14. Significant parameters in AWJM of titanium alloys based on 35 literature reviews.

 https://doi.org/10.3390/designs10010013

According to Fuse et al. [81], the optimal values of AWJM of titanium alloys are as follows: water pressure of 234.94 MPa, a stand-off distance of 1.77 mm, an abrasive flow rate of 103.41 g/min, and a traversing speed of the nozzle of 91.54 mm/min.

5.1. Abrasive Embedment

One of the key problems of AWJM is that the abrasives or parts of the abrasives embed onto the machined surfaces [76,125]. This adulteration of the surfaces causes great complications in the course of additional treatments: for example, coating, welding or grinding [126]. Embedded particles also act as initiators of fatigue cracks and reduce the fatigue strength [127]. The embedded abrasives in the Ti-6Al-4V alloy can be divided into the following: (a) larger abrasives with sizes greater than 100 μm that are close to their original size, (b) abrasive particles of 25–100 μm that are commonly found across the surfaces, and (c) microscopic abrasives with particle sizes of below 25 μm [128]. Without having any heat-affected zone components produced by AWJM, the component still failed before the laser-machined component under fatigue loading because of the ease of fatigue crack initiation as a result of grit embedment. The same was reported in [129].

Kong et al. [125]'s work on machining titanium aluminide noted "bubble-like TiO_2-based spots on the AWJ cut surfaces, which might be caused by a strong exothermic reaction". Li et al. [130] reported that the titanium surface on the Ti6Al4V/CFRP hybrid stack showed wear tracks; these are likely the traces left by repeated impacts of abrasive particles, leading to the "micro-removal of material". Fractured abrasive particles are embedded in the titanium; this indicates that not all abrasive material bounces off, as some of it becomes trapped, which can impact subsurface integrity and possibly affect strength or fatigue behaviour. Studies have reported that abrasive embodiments observed below the machined surfaces, and the number of residual abrasives and the depth of abrasives' penetration, increased with the increase in grit size, water jet velocity, and impingement angle during AWJM [131,132]. Hence, abrasive embedment can be minimised, if not eliminated, with the proper selection of parameters. Other drawbacks of AWJM include frequent replacement of nozzles due to high wear-out of materials, failure of other components of the system due to the abrasive action of the abrasive particles and the bluntness of abrasive particles after repeated usage. However, there were no works observed in the parameter optimisation studies that identified parameter combinations that will reduce abrasive embedment in Ti-6Al-4V or improve the service life of the AWJM nozzles.

5.2. Hybrid AWJM Systems

A number of works that concentrated on hybrid AWJM systems have been reported recently. Huang et al. [133] considered the "hybrid waterjet cleaning (HWJC) process on Ti-6Al-4V," where AWJM is firstly applied to remove material to a near-chosen depth, followed by WJM to remove the embedded abrasive particles and to reduce surface roughness. Though the HWJC process effectively reduced the abrasive embedment, its efficiency depends on the water jet machining (WJM)'s parameters as well as the embedment behaviour of the abrasives. The sub-surface grit elimination is not foreseen unless the plain water jet is adequately powerful to remove the material around the abrasive, and afterwards dislodge it. The lower traverse speed, which corresponds to higher energy densities, does not provide effective removal of embedded abrasives all the time. Plain WJM is more effective at grit removal when its travelling path is normal to that of AWJM compared to those in a similar path. An implanted particle might be partially submerged during AWJM, which may be partially removed by plain WJM, keeping the submerged portion on the surface because of mechanical interlocking. The fragments occupied the indentations, which make the surface flat and reduce the surface roughness. Nevertheless, if the jet

https://doi.org/10.3390/designs10010013

impulse is powerful enough, it will remove the whole particle, leaving the empty crater, which increases the surface roughness. Therefore, a distinct effect of the HWJC process on the surface roughness is not predictable.

Patel and Tandon [134] introduced "thermally enhanced abrasive water jet machining (TEAWJM)", where an oxy acetylene gas welding arrangement was used to heat and soften the workpiece locally just before machining. The MRR increased due to the combined effect of temperature and standoff distance, which increased the MRR due to the material being soft. The thermal effect also caused differences in surface morphology at the top and bottom surfaces, due to differences in the material removal phenomenon; at the bottom, "material is removed due to erosion and plastic deformation, while on the top, material removal is due to erosion by abrasive particles". Experimentation showed that the MRR increased for TEAWJM compared to AWJM for Ti-6Al-4V, as more materials were removed at a constant pressure and machining time. TEAWJM was recommended as being suitable for Ti-6Al-4V in extreme pressure and temperature environments, as the microstructural properties remained undisturbed, even when machined at higher temperatures. In addition, no cracks and residual stresses were noticed, even when the material endures thermo-mechanical quenching.

Limited work has been reported on these hybrid AWJM processes, even though they have demonstrated promising outcomes to further improve the efficiency of the AWJM of Ti-6Al-4V alloy.

6. Future Trends

Titanium alloys are relatively expensive materials and are primarily used in aerospace and biomedical applications, due to their exceptional mechanical, corrosion-resistant, and biocompatible properties. Components for these applications must be defect-free and are therefore predominantly manufactured using conventional machining processes. Although "non-traditional machining methods such as AWJM" offer several advantages over conventional processes, such as lower machining forces, the ability to produce complex geometries, and reduced tooling costs, can unpredictably alter surface and subsurface properties if not properly controlled.

Despite significant research interest in the AWJM of titanium alloys, particularly Grade 5 titanium alloy (Ti-6Al-4V), several challenges remain unresolved. These challenges mainly arise from the strong interdependence of multiple process input parameters and the large number of output responses required to define successful machining outcomes. Furthermore, achieving high surface integrity remains a major concern due to the need to simultaneously control numerous parameters, as well as issues related to abrasive embedment and severe nozzle wear caused by continuous abrasive flow. Moreover, such particle embedment can cause wear [135,136] and corrosion wear [137] in certain applications. This review has identified several key areas that warrant further investigation to enhance the reliability of AWJM for machining titanium alloys:

(a) The application of "artificial intelligence (AI) and machine learning (ML)" in AWJM process optimisation has gained notable attention, with a significant number of studies being published in 2025. However, substantial research opportunities remain, particularly in exploring alternative ML models to ANN for real-time adaptive control. Such studies should focus on accurately predicting and optimising key machining responses, including MRR, surface roughness, kerf geometry, and subsurface damage.

(b) Very limited work has been reported on understanding the relationship between AWJM process parameters and the fatigue behaviour of titanium alloys. Future research should focus on a detailed analysis of subsurface deformation, micro-cracking, and residual

https://doi.org/10.3390/designs10010013

stresses resulting from abrasive characteristics and embedment. In addition, parameter optimisation studies, using multi-objective optimisation techniques or AI-based approaches, are required to identify optimal process conditions that minimise abrasive embedment while enhancing the fatigue performance of titanium alloys.

(c) The review identified only a single study employing computational-based modelling of the AWJM process for Ti-6Al-4V. Future studies should therefore focus on developing robust numerical and computational models to analyse the influence of process parameters on jet behaviour, nozzle wear mechanisms, and resulting surface quality, thereby improving process predictability and control.

(d) The review identified only two hybrid-assisted AWJM processes. Further studies are required to develop a deeper understanding of the material removal mechanisms in hybrid-assisted AWJM and to systematically compare their performance with conventional AWJM, particularly for applications in critical industries such as aerospace and biomedical engineering.

7. Conclusions

Abrasive water jet machining (AWJM) is a non-traditional machining process that has gained considerable attention for machining titanium alloys, particularly Ti-6Al-4V, due to its ability to machine difficult-to-cut materials without inducing significant thermal damage. In AWJM, a high-velocity jet of water entrained with hard abrasive particles is employed to erode and cut the workpiece material. The process involves the conversion of pressure energy into kinetic energy, leading to material removal, primarily through plastic deformation and quasi-brittle fracture mechanisms.

AWJM is governed by a large number of process input parameters, which may be broadly classified into hydraulic parameters, nozzle parameters, abrasive parameters, and cutting parameters. These parameters collectively influence key output responses such as the material removal rate (MRR), surface roughness, kerf geometry, and subsurface integrity. An increase in water jet pressure results in higher jet kinetic energy, thereby enhancing the erosive capability of the jet and promoting material removal. The nozzle design is critical in accelerating and converging the water jet, while also facilitating efficient mixing of water and abrasive particles through the Venturi effect. The abrasive particles, possessing significantly higher hardness than the workpiece material, act as high-energy micro-cutting tools; their shape and velocity play a decisive role in the erosion and cutting mechanisms upon impact with the machined surface.

Cutting parameters, including nozzle traverse speed, standoff distance, and jet impingement angle, regulate the intensity and exposure time of the abrasive-laden jet on the workpiece surface. The water jet pressure and abrasive flow rate are generally reported to be directly proportional to the material removal rate, whereas the traverse speed and standoff distance exhibit an inverse relationship with the MRR. With respect to surface integrity, higher water jet pressure, increased traverse speed, elevated abrasive flow rate, and larger standoff distance have been shown to increase surface roughness.

Parameter optimisation studies employing experimental design methodologies and data-driven approaches, such as artificial neural networks, consistently identify water jet pressure and nozzle traverse speed as the most influential parameters governing surface quality in AWJM of titanium alloys. Despite the extensive body of literature on process parameter optimisation, limited attention has been paid to abrasive embedment on the machined surface and the development of hybrid AWJM techniques. Accordingly, this review highlights these research gaps and proposes potential future research directions to advance the understanding and industrial applicability of AWJM for titanium alloy machining.

 https://doi.org/10.3390/designs10010013

Author Contributions: A.A.: Resource, Data curation, Writing, Writing—review and editing; A.P.: Conceptualization, Validation, Writing, Writing—review and editing; A.R.D.: Formal analysis, Validation, Data curation, Writing—review and editing, A.K.B.: Conceptualization, Writing, Writing—review and editing, Supervision. All authors have read and agreed to the published version of the manuscript.

Funding: This research received no external funding.

Data Availability Statement: The raw/processed data used to produce the results will be made available by the corresponding author upon reasonable request.

Conflicts of Interest: The authors declare that they have no known competing financial interest or personal relationship whatsoever that could have influenced the work reported in this paper.

References

1. Pushp, P.; Dasharath, S.M.; Arati, C. Classification and applications of titanium and its alloys. *Mater. Today Proc.* **2022**, *54*, 537–542. [CrossRef]
2. Arrazola, P.J.; Garay, A.; Iriarte, L.M.; Armendia, M.; Marya, S.; Le Maître, F. Machinability of titanium alloys (Ti6Al4V and Ti555.3). *J. Mater. Process. Technol.* **2009**, *209*, 2223–2230. [CrossRef]
3. Biffi, C.A.; Lecis, N.; Previtali, B.; Vedani, M.; Vimercati, G.M. Fiber laser microdrilling of titanium and its effect on material microstructure. *Int. J. Adv. Manuf. Technol.* **2011**, *54*, 149–160. [CrossRef]
4. Bammidi, R.; Sreeramulu, D.; Madivada, H.; Rejeti, P.K.; Venkatesh, M. Towards an understanding of Ti-6Al-4V machining and machinability. *Mater. Today Proc.* **2024**, *115*, 289–292. [CrossRef]
5. Pramanik, A.; Littlefair, G. Developments in machining of stacked materials made of CFRP and titanium/aluminum alloys. *Mach. Sci. Technol.* **2014**, *18*, 485–508. [CrossRef]
6. Pramanik, A. Developments in the non-traditional machining of particle reinforced metal matrix composites. *Int. J. Mach. Tools Manuf.* **2014**, *86*, 44–61. [CrossRef]
7. Pramanik, A.; Neo, K.; Rahman, M.; Li, X.; Sawa, M.; Maeda, Y. Ultraprecision turning of electroless nickel: Effects of crystal orientation and origin of diamond tools. *Int. J. Adv. Manuf. Technol.* **2009**, *43*, 681. [CrossRef]
8. Kim, D.-H.; Lee, C.-M. Experimental Investigation on Machinability of Titanium Alloy by Laser-Assisted End Milling. *Metals* **2021**, *11*, 1552. [CrossRef]
9. Pramanik, A. Problems and solutions in machining of titanium alloys. *Int. J. Adv. Manuf. Technol.* **2014**, *70*, 919–928. [CrossRef]
10. Pramanik, A.; Littlefair, G. Machining of titanium alloy (Ti-6Al-4V)—Theory to application. *Mach. Sci. Technol.* **2015**, *19*, 1–49. [CrossRef]
11. Donachie, M.J. *Titanium: A Technical Guide*; ASM International: Materials Park, OH, USA, 2000.
12. Pandey, A.K.; Dubey, A.K. Simultaneous optimization of multiple quality characteristics in laser cutting of titanium alloy sheet. *Opt. Laser Technol.* **2012**, *44*, 1858–1865. [CrossRef]
13. Pal, V.K.; Tandon, P. Identification of the role of machinability and milling depth on machining time in controlled depth milling using abrasive water jet. *Int. J. Adv. Manuf. Technol.* **2013**, *66*, 877–881. [CrossRef]
14. Kumar, J. *Investigating the Machining Characteristics of Titanium Using Ultrasonic Machining*; Thapar Institute of Engineering Technology: Patiala, India, 2009.
15. Patil Amit, S.; Nashik, I.S.V.; More Yogesh, S.; Nathe Manik, S. Machining Challenges in Ti-6Al-4V.-A Review. *Int. J. Innov. Eng. Technol.* **2015**, *5*, 17.
16. Mruthunjaya, M.; Yogesha, K.B. A review on conventional and thermal assisted machining of titanium based alloy. *Mater. Today Proc.* **2021**, *46*, 8466–8472. [CrossRef]
17. Oke, S.R.; Ogunwande, G.S.; Onifade, M.; Aikulola, E.; Adewale, E.D.; Olawale, O.E.; Ayodele, B.E.; Mwema, F.; Obiko, J.; Bodunrin, M.O. An overview of conventional and non-conventional techniques for machining of titanium alloys. *Manuf. Rev.* **2020**, *7*, 34. [CrossRef]
18. Nouari, M.; Makich, H. On the Physics of Machining Titanium Alloys: Interactions between Cutting Parameters, Microstructure and Tool Wear. *Metals* **2014**, *4*, 335–358. [CrossRef]
19. Yuan, C.G.; Pramanik, A.; Basak, A.K.; Prakash, C.; Shankar, S. Drilling of titanium alloy (Ti6Al4V)—A review. *Mach. Sci. Technol.* **2021**, *25*, 637–702. [CrossRef]
20. Basak, A.; Pramanik, A.; Prakash, C.; Shankar, S.; Debnath, S. Understanding the micro-mechanical behaviour of recast layer formed during WEDM of titanium alloy. *Metals* **2022**, *12*, 188. [CrossRef]

 https://doi.org/10.3390/designs10010013

21. Alshemary, A.; Pramanik, A.; Basak, A.; Littlefair, G. Accuracy of duplex stainless steel feature generated by electrical discharge machining (EDM). *Measurement* **2018**, *130*, 137–144. [CrossRef]
22. Rajurkar, K.P.; Zhu, D.; McGeough, J.; Kozak, J.; De Silva, A. New developments in electro-chemical machining. *CIRP Ann.* **1999**, *48*, 567–579. [CrossRef]
23. Pramanik, A.; Basak, A.K. Laser beam machining of titanium alloy—A review. *Metals* **2023**, *13*, 1536. [CrossRef]
24. Mardi, K.B.; Dixit, A.; Pramanik, A.; Basak, A. Tribology in (abrasive) water jet machining: A review. *Mach. Tribol.* **2022**, 113–125. [CrossRef]
25. Natarajan, Y.; Murugesan, P.K.; Mohan, M.; Khan, S.A.L.A. Abrasive Water Jet Machining process: A state of art of review. *J. Manuf. Process.* **2020**, *49*, 271–322. [CrossRef]
26. Pramanik, A.; Littlefair, G. Wire EDM mechanism of MMCs with the variation of reinforced particle size. *Mater. Manuf. Process.* **2016**, *31*, 1700–1708. [CrossRef]
27. Mandal, A.; Dixit, A.R.; Chattopadhyaya, S.; Paramanik, A.; Hloch, S.; Królczyk, G. Improvement of surface integrity of Nimonic C 263 super alloy produced by WEDM through various post-processing techniques. *Int. J. Adv. Manuf. Technol.* **2017**, *93*, 433–443. [CrossRef]
28. Wang, F.; Liu, Y.; Zhang, Y.; Tang, Z.; Ji, R.; Zheng, C. Compound machining of titanium alloy by super high speed EDM milling and arc machining. *J. Mater. Process. Technol.* **2014**, *214*, 531–538. [CrossRef]
29. Chang, S.; Liu, W.; Tang, J.; Fan, M. Thermomechanical coupling mechanisms and surface integrity in the abrasive waterjet machining of Ti-5Al-5Mo-5V-1Fe-1Cr alloy. *Mater. Today Commun.* **2025**, *46*, 112770. [CrossRef]
30. Hashish, M. Characteristics of surfaces machined with abrasive-waterjets. *J. Eng. Mater. Technol.* **1991**, *113*, 354–362. [CrossRef]
31. Hashish, M. An investigation of milling with abrasive-waterjets. *J. Eng. Ind.* **1989**, *111*, 158–166. [CrossRef]
32. Ozcan, Y.; Tunc Lutfi, T.; Kopacka, J.; Cetin, B.; Sulitka, M. Modelling and simulation of controlled depth abrasive water jet machining (AWJM) for roughing passes of free-form surfaces. *Int. J. Adv. Manuf. Technol.* **2021**, *114*, 3581–3596. [CrossRef]
33. Momber, A.W.; Kovacevic, R. *Principles of Abrasive Water Jet Machining*; Springer Science & Business Media: Berlin/Heidelberg, Germany, 2012.
34. Finnie, I. Erosion of surfaces by solid particles. *Wear* **1960**, *3*, 87–103. [CrossRef]
35. Summers, D.A. *Waterjetting Technology*; CRC Press: Boca Raton, FL, USA, 2003.
36. Guo, Z.; Ramulu, M. Investigation of displacement fields in an abrasive waterjet drilling process: Part 1. Experimental measurements. *Exp. Mech.* **2001**, *41*, 375–387. [CrossRef]
37. Li, H.; Wang, J. An experimental study of abrasive waterjet machining of Ti-6Al-4V. *Int. J. Adv. Manuf. Technol.* **2015**, *81*, 361–369. [CrossRef]
38. Basak, A.; Fan, J.; Wang, J.; Mathew, P. Material removal mechanisms of monocrystalline silicon under the impact of high velocity micro-particles. *Wear* **2010**, *269*, 269–277. [CrossRef]
39. Wang, H.; Yuan, R.; Zhang, X.; Zai, P.; Deng, J. Research Progress in Abrasive Water Jet Processing Technology. *Micromachines* **2023**, *14*, 1526. [CrossRef]
40. Wang, J.; Wong, W. A study of abrasive waterjet cutting of metallic coated sheet steels. *Int. J. Mach. Tools Manuf.* **1999**, *39*, 855–870. [CrossRef]
41. Miao, X.; Wu, M. Modeling of cutting of stainless steel AISI 304 by abrasive water jet. *Mater. Res. Express* **2020**, *7*, 086507. [CrossRef]
42. Hashish, M. Abrasive Waterjet Machining. *Materials* **2024**, *17*, 3273. [CrossRef]
43. Hlaváč, L.M. Revised Model of Abrasive Water Jet Cutting for Industrial Use. *Materials* **2021**, *14*, 4032. [CrossRef]
44. Kong, M.; Axinte, D.; Voice, W. Aspects of material removal mechanism in plain waterjet milling on gamma titanium aluminide. *J. Mater. Process. Technol.* **2010**, *210*, 573–584. [CrossRef]
45. Hashish, M. Optimization factors in abrasive-waterjet machining. *J. Eng. Ind.* **1991**, *113*, 29–37. [CrossRef]
46. Ramakrishnan, S. Investigating the effects of abrasive water jet machining parameters on surface integrity, chemical state in machining of Ti-6Al-4V. *Mater. Today Commun.* **2022**, *31*, 103480. [CrossRef]
47. de la Rosa, S.; Rodríguez-Parada, L.; Ponce, M.B.; Mayuet Ares, P.F. The Abrasive Water Jet Cutting Process of Carbon-Fiber-Reinforced Polylactic Acid Samples Obtained by Additive Manufacturing: A Comparative Analysis. *J. Compos. Sci.* **2024**, *8*, 437. [CrossRef]
48. Chaturvedi, C.; Kumar, H.; Nagdeve, L. Machining Efficiency and Property Change Analysis of Ti-6Al-4V Alloy Processed by Abrasive Water Jet Machining. *J. Mater. Eng. Perform.* **2024**, *34*, 8551–8561. [CrossRef]
49. Ramakrishnan, S.; Singaravelu, D.L.; Senthilkumar, V.; Velmurugan, C. Experimental investigation on microstructural characterization of machined Ti-6Al-4V using abrasive water jet machining. *Sadhana* **2022**, *47*, 83. [CrossRef]

50. Kanthababu, M.; Ram, R.; Emannuel, N.; Gokul, R.; Rammohan, R. Experimental investigations on pocket milling of titanium alloy using abrasive water jet machining. *FME Trans.* **2016**, *44*, 133–138. [CrossRef]
51. Natarajan, Y.; Raj, K.L.N.; Tandon, P. Measurement and analysis of pocket milling features in abrasive water jet machining of Ti-6Al-4V alloy. *Arch. Civ. Mech. Eng.* **2023**, *23*, 42. [CrossRef]
52. Liu, H. Precision machining of advanced materials with waterjets. *IOP Conf. Ser. Mater. Sci. Eng.* **2017**, *164*, 012008. [CrossRef]
53. Wu, L.; Zou, X.; Guo, Y.; Fu, L. Parametric Structure Optimization Design of High-Pressure Abrasive Water Jet Nozzle Based on Computational Fluid Dynamics-Discrete Element Method (CFD-DEM). *Lubricants* **2025**, *13*, 91. [CrossRef]
54. Llanto, J.M.; Vafadar, A.; Aamir, M.; Tolouei-Rad, M. Analysis and Optimization of Process Parameters in Abrasive Waterjet Contour Cutting of AISI 304L. *Metals* **2021**, *11*, 1362. [CrossRef]
55. Pătîrnac, I.; Ripeanu, R.G.; Popescu, C.; Tănase, M. Analysis of Water Flow through the Active Parts of an Abrasive Water Jet Machine: A Combined Analytical and CFD Approach. *Processes* **2024**, *12*, 1551. [CrossRef]
56. Pătîrnac, I.; Ripeanu, R.G.; Tănase, M. Computational Fluid Dynamics Analysis of Erosion in Active Components of Abrasive Water Jet Machine. *Processes* **2024**, *12*, 1860. [CrossRef]
57. Fowler, G.; Pashby, I.; Shipway, P. The effect of particle hardness and shape when abrasive water jet milling titanium alloy Ti6Al4V. *Wear* **2009**, *266*, 613–620. [CrossRef]
58. Boud, F.; Carpenter, C.; Folkes, J.; Shipway, P. Abrasive waterjet cutting of a titanium alloy: The influence of abrasive morphology and mechanical properties on workpiece grit embedment and cut quality. *J. Mater. Process. Technol.* **2010**, *210*, 2197–2205. [CrossRef]
59. Karkalos, N.E.; Dekster, L.; Kudelski, R.; Karmiris-Obratański, P. A Statistical and Optimization Study on the Influence of Different Abrasive Types on Kerf Quality and Productivity during Abrasive Waterjet (AWJ) Milling of Ti-4Al-6V. *Materials* **2023**, *17*, 11. [CrossRef] [PubMed]
60. Mort, G.A. Results of abrasive water jet market survey. In Proceedings of the of 8th American Water Jet Conference, Houston, TX, USA, 26–29 August 1995.
61. Babu, M.K.; Chetty, O.K. Studies on recharging of abrasives in abrasive water jet machining. *Int. J. Adv. Manuf. Technol.* **2002**, *19*, 697–703. [CrossRef]
62. Fowler, G.; Shipway, P.; Pashby, I. A technical note on grit embedment following abrasive water-jet milling of a titanium alloy. *J. Mater. Process. Technol.* **2005**, *159*, 356–368. [CrossRef]
63. Hascalik, A.; Çaydaş, U.; Gürün, H. Effect of traverse speed on abrasive waterjet machining of Ti–6Al–4V alloy. *Mater. Des.* **2007**, *28*, 1953–1957. [CrossRef]
64. Karkalos, N.E.; Karmiris-Obratański, P. Determination of the Feasibility of Using Eco-Friendly Walnut Shell Abrasive Particles for Pocket Milling of Titanium Workpieces by Abrasive Waterjet Technology. *Metals* **2023**, *13*, 1645. [CrossRef]
65. Muzammil, S.B.; Endale, A.E.; Gasha, S.B. Studies on machining characteristics of titanium alloy Ti-6Al-4V using abrasive water jet cutting. *J. Eng. Res. Appl.* **2018**, *8*, 7.
66. Perec, A. Experimental research into alternative abrasive material for the abrasive water-jet cutting of titanium. *Int. J. Adv. Manuf. Technol.* **2018**, *97*, 1529–1540. [CrossRef]
67. Fengchao, W.; Kaining, Y.; Nanzhe, X.; Liucan, L.; Chuwen, G. Quantitative analysis on the surface topography of Ti-6Al-4V machined by abrasive suspension jet. *Int. J. Adv. Manuf. Technol.* **2021**, *112*, 323–332. [CrossRef]
68. Umanath, K.; Devika, D.; Begum, R.S. Experimental Investigation of Role of Particles Size and Cutting Passes in Abrasive Waterjet Machining Process on Titanium Alloy (Ti–6Al–4V) using Taguchi's Method. *Mater. Manuf. Process.* **2021**, *36*, 936–949. [CrossRef]
69. Sami Abushanab, W.; Moustafa, E.B.; Harish, M.; Shanmugan, S.; Elsheikh, A.H. Experimental investigation on surface characteristics of Ti6Al4V alloy during abrasive water jet machining process. *Alex. Eng. J.* **2022**, *61*, 7529–7539. [CrossRef]
70. Kumar, N.; Shukla, M. Finite element analysis of multi-particle impact on erosion in abrasive water jet machining of titanium alloy. *J. Comput. Appl. Math.* **2012**, *236*, 4600–4610. [CrossRef]
71. Shipway, P.; Fowler, G.; Pashby, I. Characteristics of the surface of a titanium alloy following milling with abrasive waterjets. *Wear* **2005**, *258*, 123–132. [CrossRef]
72. Sharma, V.S.; Kumar, A.; Gupta, M.K.; Bhanot, N. *Process Parameter Optimization for Abrasive Water Jet Machining of Titanium Alloy Using Meta-Heuristic Algorithms*; EDP Sciences: Les Ulis, France.
73. Gnanavelbabu, A.; Saravanan, P.; Rajkumar, K.; Karthikeyan, S. Experimental Investigations on Multiple Responses in Abrasive Waterjet Machining of Ti-6Al-4V Alloy. *Mater. Today Proc.* **2018**, *5*, 13413–13421. [CrossRef]
74. Fuse, K.; Chaudhari, R.; Vora, J.; Patel, V.K.; de Lacalle, L.N.L. Multi-Response Optimization of Abrasive Waterjet Machining of Ti6Al4V Using Integrated Approach of Utilized Heat Transfer Search Algorithm and RSM. *Materials* **2021**, *14*, 7746. [CrossRef]

75. Chithirai Pon Selvan, M.; Midhunchakkaravarthy, D.; Senanayake, R.; Ramaswamy Pillai, S.; Reddy Madara, S. A mathematical modelling of Abrasive Waterjet Machining on Ti-6Al-4V using Artificial Neural Network. *Mater. Today Proc.* **2020**, *28*, 538–544. [CrossRef]
76. Seo, Y.; Ramulu, M.; Kim, D. Machinability of titanium alloy (Ti'6Al'4V) by abrasive waterjets. *Proc. Inst. Mech. Eng. Part B J. Eng. Manuf.* **2003**, *217*, 1709–1721. [CrossRef]
77. Wang, J. Abrasive waterjet machining of polymer matrix composites–cutting performance, erosive process and predictive models. *Int. J. Adv. Manuf. Technol.* **1999**, *15*, 757–768. [CrossRef]
78. Hlaváč, L.M.; Gembalová, L.; Štěpán, P.; Hlaváčová, I.M. Improvement of abrasive water jet machining accuracy for titanium and TiNb alloy. *Int. J. Adv. Manuf. Technol.* **2015**, *80*, 1733–1740. [CrossRef]
79. Arola, D.; Ramulu, M. Material removal in abrasive waterjet machining of metals surface integrity and texture. *Wear* **1997**, *210*, 50–58. [CrossRef]
80. Ruiz-Garcia, R.; Mayuet Ares, P.F.; Vazquez-Martinez, J.M.; Salguero Gómez, J. Influence of abrasive waterjet parameters on the cutting and drilling of CFRP/UNS A97075 and UNS A97075/CFRP stacks. *Materials* **2018**, *12*, 107. [CrossRef] [PubMed]
81. Fuse, K.; Vora, J.; Wakchaure, K.; Patel, V.K.; Chaudhari, R.; Saxena, K.K.; Bandhu, D.; Ramacharyulu, D.A. Abrasive waterjet machining of titanium alloy using an integrated approach of taguchi-based passing vehicle search algorithm. *Int. J. Interact. Des. Manuf.* **2025**, *19*, 2249–2263. [CrossRef]
82. Dubey, A.K.; Kumar, Y.; Kumar, S.; Raja, A.R. Parametric Optimization of AWJM Using RSM-Grey-TLBO-Based MCDM Approach for Titanium Grade 5 Alloy. *Arab. J. Sci. Eng.* **2025**, *50*, 13693–13711. [CrossRef]
83. Ramakrishnan, S.; Senthilkumar, V.; Lenin Singaravelu, D. Effect of cutting parameters on surface integrity characteristics of Ti-6Al-4V in abrasive water jet machining process. *Mater. Res. Express* **2019**, *6*, 116583. [CrossRef]
84. Sultan, T.; Gilles, P.; Cohen, G.; Cenac, F.; Rubio, W. Modeling incision profile in AWJM of Titanium alloys Ti6Al4V. *Mech. Ind.* **2016**, *17*, 403. [CrossRef]
85. Karkalos, N.E.; Karmiris-Obratański, P. A Comparative Study of Efficient Modeling Approaches for Performing Controlled-Depth Abrasive Waterjet Pocket Milling. *Machines* **2024**, *12*, 168. [CrossRef]
86. Xiong, J.; Wan, L.; Qian, Y.n.; Sun, S.; Li, D.; Wu, S. A new strategy for improving the surface quality of Ti6Al4V machined by abrasive water jet: Reverse cutting with variable standoff distances. *Int. J. Adv. Manuf. Technol.* **2022**, *120*, 5339–5350. [CrossRef]
87. Van Hung, B.; Gilles, P.; Cohen, G.; Rubio, W. Develop Model for Controlled Depth Milling by Abrasive Water Jet of Ti6Al4V at Jet Inclination Angle. In *International Joint Conference on Mechanics, Design Engineering & Advanced Manufacturing*; Springer: Cham, Switzerland, 2021; pp. 21–27.
88. Dekster, L.; Karkalos, N.E.; Karmiris-Obratański, P.; Markopoulos, A.P. Evaluation of the Machinability of Ti-6Al-4V Titanium Alloy by AWJM Using a Multipass Strategy. *Appl. Sci.* **2023**, *13*, 3774. [CrossRef]
89. Pan, H.; Chen, X.; Chen, L.; You, H.; Liang, X. Surface Topography Analysis of BK7 with Different Roughness Nozzles Using an Abrasive Water Jet. *Materials* **2024**, *17*, 4494. [CrossRef] [PubMed]
90. Vasanth, S.; Muthuramalingam, T.; Vinothkumar, P.; Geethapriyan, T.; Murali, G. Performance Analysis of Process Parameters on Machining Titanium (Ti-6Al-4V) Alloy Using Abrasive Water Jet Machining Process. *Procedia CIRP* **2016**, *46*, 139–142. [CrossRef]
91. Hloch, S.; Valíček, J. Prediction of distribution relationship of titanium surface topography created by abrasive waterjet. *Int. J. Surf. Sci. Eng.* **2011**, *5*, 152–168. [CrossRef]
92. Monková, K.; Monka, P.; Vegnerová, P.; Čep, R.; Müllerová, J.; Bražina, D.; Duspara, M. Factor analysis of the abrasive waterjet factors affecting the surface roughness of titanium. *Teh. Vjesn.* **2011**, *18*, 73–77.
93. Chen, F.L.; Wang, J.; Lemma, E.; Siores, E. Striation formation mechanisms on the jet cutting surface. *J. Mater. Process. Technol.* **2003**, *141*, 213–218. [CrossRef]
94. Lenin Raj, S.; Rajadurai, A. Experimental study on deep-hole making in Ti-6Al-4V by abrasive water jet machining. *Mater. Res. Express* **2019**, *6*, 66532. [CrossRef]
95. Chaturvedi, C.; Rao, P.S.; Khan, M.Y. Optimization of process variable in abrasive water jet Machining (AWJM) of Ti-6Al-4V alloy using Taguchi methodology. *Mater. Today Proc.* **2021**, *47*, 6120–6127. [CrossRef]
96. Ahmed, T.M.; El Mesalamy, A.S.; Youssef, A.; El Midany, T.T. Improving surface roughness of abrasive waterjet cutting process by using statistical modeling. *CIRP J. Manuf. Sci. Technol.* **2018**, *22*, 30–36. [CrossRef]
97. Soori, M.; Arezoo, B. Minimization of surface roughness and residual stress in abrasive water jet cutting of titanium alloy Ti6Al4V. *Proc. Inst. Mech. Eng. Part E J. Process Mech. Eng.* **2024**, *238*, 1613–1625. [CrossRef]
98. Thirumalai Kumaran, S.; Uthayakumar, M.; Mathiyazhagan, P.; Krishna Kumar, K.; Muthu Kumar, P. Effect of abrasive grain size of the AWJM performance on AA (6351)-SiC-B4C hybrid composite. *Appl. Mech. Mater.* **2015**, *766*, 324–329. [CrossRef]
99. Mohamad, W.; Kasim, M.; Norazlina, M.; Hafiz, M.; Izamshah, R.; Mohamed, S. Effect of standoff distance on the kerf characteristic during abrasive water jet machining. *Results Eng.* **2020**, *6*, 100101. [CrossRef]

100. Ramulu, M.; Arola, D. Water jet and abrasive water jet cutting of unidirectional graphite/epoxy composite. *Composites* **1993**, *24*, 299–308. [CrossRef]
101. Nagendra Prasad, K.; John Basha, D.; Varaprasad, K.C. Experimental Investigation and Analysis of Process Parameters in Abrasive Jet Machining of Ti-6Al-4V alloy using Taguchi Method. *Mater. Today Proc.* **2017**, *4*, 10894–10903. [CrossRef]
102. Muthuramalingam, T.; Vasanth, S.; Vinothkumar, P.; Geethapriyan, T.; Rabik, M.M. Multi Criteria Decision Making of Abrasive Flow Oriented Process Parameters in Abrasive Water Jet Machining Using Taguchi–DEAR Methodology. *Silicon* **2018**, *10*, 2015–2021. [CrossRef]
103. Mogul, Y.I.; Nasir, I.; Myler, P. Investigation and optimization for depth of cut and surface roughness for control depth milling in Titanium Ti6AL4V with abrasive water jet cutting. *Mater. Today Proc.* **2020**, *28*, 604–610. [CrossRef]
104. Karkalos, N.E.; Karmiris-Obratański, P.; Kudelski, R.; Markopoulos, A.P. Experimental Study on the Sustainability Assessment of AWJ Machining of Ti-6Al-4V Using Glass Beads Abrasive Particles. *Sustainability* **2021**, *13*, 8917. [CrossRef]
105. K, S.; J., F.X.; M.P, S.; Maridurai, T.; Suyamburajan, V.; Jayaseelan, V. Optimization of SiC Abrasive Parameters on Machining of Ti-6Al-4V Alloy in AJM Using Taguchi-Grey Relational Method. *Silicon* **2022**, *14*, 997–1004. [CrossRef]
106. Perumal, A.; Kailasanathan, C.; Wilson, V.H.; Sampath Kumar, T.; Stalin, B.; Rajkumar, P.R. Machinability of Titanium alloy 6242 by AWJM through Taguchi method. *Mater. Today Proc.* **2023**, *81*, 606–611. [CrossRef]
107. Wan, L.; Liu, J.; Qian, Y.n.; Wang, X.; Wu, S.; Du, H.; Li, D. Analytical modeling and multi-objective optimization algorithm for abrasive waterjet milling Ti6Al4V. *Int. J. Adv. Manuf. Technol.* **2022**, *123*, 4367–4384. [CrossRef]
108. Radomska-Zalas, A. Experimental Research on the Use of a Selected Multi-Criteria Method for the Cutting of Titanium Alloy with an Abrasive Water Jet. *Materials* **2023**, *16*, 5405. [CrossRef]
109. Tomy, A.; Hiremath, S.S. An experimental investigation and multi-objective optimization of abrasive jet machining on Ti-6Al-4V ELI bio-material. *J. Micromanuf.* **2023**, *6*, 50–59. [CrossRef]
110. Yuan, Y.; Chen, J.; Gao, H.; Wang, X. An investigation into the abrasive waterjet milling circular pocket on titanium alloy. *Int. J. Adv. Manuf. Technol.* **2020**, *107*, 4503–4515. [CrossRef]
111. Dubey, A.K.; Kumar, Y.; Kumar, S. Optimizing Parameters of AWJM for Ti-6Al-4 V Grade 5 Alloy Using Grey Entropy Weight Method: A Multivariable Approach. *J. Inst. Eng. India Ser. C* **2025**, *106*, 181–195. [CrossRef]
112. Kumar, R.; Krishna, M.G.; Billu, E.; Reddy, D.A.K.; Ram, S. Optimization of surface roughness in machining of titanium composite using taguchi method. *Interactions* **2025**, *246*, 33. [CrossRef]
113. Naresh Raj, K.L.; Jayavelu, S.; Kumar, M.S.; Kumar, A.; Swarna, B. Multi-criteria decision analysis approach on parametric optimization of abrasive waterjet pocket milling in Ti-6Al-4V alloy. *Sci. Rep.* **2025**, *15*, 42719. [CrossRef]
114. Popan, I.A.; Bocăneț, V.I.; Softic, S.; Popan, A.I.; Panc, N.; Balc, N. Artificial Intelligence Model Used for Optimizing Abrasive Water Jet Machining Parameters to Minimize Delamination in Carbon Fiber-Reinforced Polymer. *Appl. Sci.* **2024**, *14*, 8512. [CrossRef]
115. Periyappillai, G.; Subbarayan, S.; Sengottaiyan, S. Advanced ensemble machine learning prediction to enhance the accuracy of abrasive waterjet machining for biocomposites. *Mater. Chem. Phys.* **2025**, *333*, 130175. [CrossRef]
116. Ficko, M.; Begic-Hajdarevic, D.; Cohodar Husic, M.; Berus, L.; Cekic, A.; Klancnik, S. Prediction of Surface Roughness of an Abrasive Water Jet Cut Using an Artificial Neural Network. *Materials* **2021**, *14*, 3108. [CrossRef]
117. Doreswamy, D.; Bhat, S.K.; Hiremath, P.; Shreyas, D.S.; Bongale, A. ANN-based predictive modelling of the effect of abrasive water-jet parameters on the surface roughness of AZ31 Mg alloy. *Manuf. Rev.* **2024**, *11*, 21.
118. Ting, H.Y.; Asmelash, M.; Azhari, A.; Alemu, T.; Saptaji, K. Prediction of surface roughness of titanium alloy in abrasive waterjet machining process. *Int. J. Interact. Des. Manuf.* **2022**, *16*, 281–289. [CrossRef]
119. Abouhawa, M.; Elshinawi, A.; Eisa, A.; Fattouh, M. A hybrid machine learning approach to optimize kerf quality in AWJM of Ti-6Al-4V/CFRP/Al7075 stacks. *Int. J. Interact. Des. Manuf.* **2025**, *19*, 7413–7435. [CrossRef]
120. Kumar, T.S.K.; Kaviti, A.K. An advanced study on the application of artificial neural networks in the abrasive waterjet machining of titanium nanocomposites. *Int. J. Interact. Des. Manuf.* **2025**, *19*, 4463–4470. [CrossRef]
121. Thakur, P.; Khemchandani, M.; Deshmukh, M. Statistical and Artificial Intelligence-Based Modeling of Multi-Parameter Evaluation Index (MPEI) in Abrasive Waterjet Cutting Process. *Arab. J. Sci. Eng.* **2025**, *50*, 18799–18814. [CrossRef]
122. Jithendra, T.; Basha, S.S. A forecasting model for optimizing abrasive water jet machining (AWJM) parameters based on the adaptive neuro-fuzzy inference system and meta-heuristic algorithms. *Proc. Inst. Mech. Eng. Part E J. Process Mech. Eng.* **2025**, *239*, 3197–3210. [CrossRef]
123. Mogul, Y.I.; Mogul, I.; Quadros, J.D.; Mohin, M.; Aabid, A.; Baig, M.; Malik, M.A. Prediction of cutting depth in abrasive water jet machining of Ti-6AL-4V alloy using back propagation neural networks. *Results Eng.* **2025**, *25*, 104520. [CrossRef]

 https://doi.org/10.3390/designs10010013

124. Parthasarathi, S.; Induja, I.; Anila, S.; Nisha, D.; Prija, V.; Hansen, R. Optimization of Machining Parameters for Ti-6Al-4V Nano-alloy in Abrasive Jet Machining Using Taguchi-Grey Relational Analysis with ANN. *J. Environ. Nanotechnol.* **2025**, *14*, 477–485. [CrossRef]
125. Kong, M.; Axinte, D. Response of titanium aluminide alloy to abrasive waterjet cutting: Geometrical accuracy and surface integrity issues versus process parameters. *Proc. Inst. Mech. Eng. Part B J. Eng. Manuf.* **2009**, *223*, 19–42. [CrossRef]
126. Chen, F.L.; Siores, E.; Patel, K.; Momber, A. Minimising particle contamination at abrasive waterjet machined surfaces by a nozzle oscillation technique. *Int. J. Mach. Tools Manuf.* **2002**, *42*, 1385–1390. [CrossRef]
127. Singh, J.; Jain, S.C. Mechanical issues in laser and abrasive water jet cutting. *JOM* **1995**, *47*, 28–30. [CrossRef]
128. Sourd, X.; Zitoune, R.; Hejjaji, A.; Salem, M.; Hor, A.; Lamouche, D. Plain water jet cleaning of titanium alloy after abrasive water jet milling: Surface contamination and quality analysis in the context of maintenance. *Wear* **2021**, *477*, 203833. [CrossRef]
129. Fordham, J.; Pilkington, R.; Tang, C. The effect of different profiling techniques on the fatigue performance of metallic membranes of AISI 301 and Inconel 718. *Int. J. Fatigue* **1997**, *6*, 487–501.
130. Li, M.; Huang, M.; Chen, Y.; Kai, W.; Yang, X. Experimental study on hole characteristics and surface integrity following abrasive waterjet drilling of Ti6Al4V/CFRP hybrid stacks. *Int. J. Adv. Manuf. Technol.* **2019**, *104*, 4779–4789. [CrossRef]
131. Maruyama, T.; Akagi, K.; Kobayashi, T. Effects of blasting parameters on removability of residual grit. *J. Therm. Spray Technol.* **2006**, *15*, 817–821. [CrossRef]
132. Arola, D.; Hall, C.L. Parametric Effects on Particle Deposition in Abrasive Waterjet Surface Treatments. *Mach. Sci. Technol.* **2004**, *8*, 171–192. [CrossRef]
133. Huang, L.; Kinnell, P.; Shipway, P. Parametric effects on grit embedment and surface morphology in an innovative hybrid waterjet cleaning process for alpha case removal from titanium alloys. *Procedia CIRP* **2013**, *6*, 594–599. [CrossRef]
134. Patel, D.; Tandon, P. Experimental investigations of thermally enhanced abrasive water jet machining of hard-to-machine metals. *CIRP J. Manuf. Sci. Technol.* **2015**, *10*, 92–101. [CrossRef]
135. Basak, A.K.; Celis, J.-P.; Vardavoulias, M.; Matteazzi, P. Effect of nanostructuring and Al alloying on friction and wear behaviour of thermal sprayed WC–Co coatings. *Surf. Coat. Technol.* **2012**, *206*, 3508–3516. [CrossRef]
136. Basak, A.; Celis, J.-P.; Vardavoulias, M.; Matteazzi, P. Abrasive wear of nanostructured cermet coatings in dry and slurry conditions. *Int. J. Refract. Met. Hard Mater.* **2021**, *100*, 105638. [CrossRef]
137. Basak, A.; Matteazzi, P.; Vardavoulias, M.; Celis, J.-P. Corrosion–wear behaviour of thermal sprayed nanostructured FeCu/WC–Co coatings. *Wear* **2006**, *261*, 1042–1050. [CrossRef]

designs

MDPI

Article

Investigation on the Microstructure and Micro-Mechanical Properties of Thermal-Sprayed NiCoCrAlY High Entropy Alloy Coating

Animesh Kumar Basak [1,2,*], Nachimuthu Radhika [3], Chander Prakash [4] and Alokesh Pramanik [4,5]

1 Adelaide Microscopy, The University of Adelaide, Adelaide, SA 5005, Australia
2 Chitkara Centre for Research and Development, Chitkara University, Baddi 174103, Himachal Pradesh, India
3 Department of Mechanical Engineering, Amrita School of Engineering, Amrita Vishwa Vidyapeetham, Coimbatore 641112, Tamil Nadu, India; n_radhika1@cb.amrita.edu
4 Centre for Research Impact & Outcome, Chitkara University, Rajpura 140401, Punjab, India; chander.mechengg@gmail.com (C.P.); alokesh.pramanik@curtin.edu.au (A.P.)
5 School of Civil and Mechanical Engineering, Curtin University, Bentley, WA 6102, Australia
* Correspondence: animesh.basak@adelaide.edu.au

Abstract: NiCoCrAlY high entropy alloy (HEA) coating (47.1 wt.% Ni, 23 wt.% Co, 17 wt.% Cr, 12.5 wt.% Al, and 0.4 wt.% Y) was deposited on a stainless steel subtract by atmospheric plasma spraying (APS). The as-deposited coating was about 300 μm thickness with <1% porosity. The microstructure of the coating consisted of dispersed secondary phases/intermetallics in the solid solution. The stress–strain behaviour of this coating was investigated in micro-scale with the help of in situ micro-pillar compression. The experimental results show that yield and compressive stress in the cross-section of the coating was higher (1.27 ± 0.10 MPa and 2.19 ± 0.10 GPa, respectively) than that of the planar direction (0.85 ± 0.09 MPa and 1.20 ± 0.08 GPa, respectively). The various secondary/intermetallic phases (γ'–Ni_3Al, β–NiAl) that were present in the coating microstructure hinder the lattice movement during compression, according to Orowan mechanism. In addition to that, the direction of the loading to that of the orientation of the phase/splat boundaries dictate the crack propagation architecture, which results in difference in the micro-mechanical properties.

Keywords: NiCoCrAlY; high entropy alloy (HEA); coating; micro-pillar compression; thermal spraying; deformation

1. Introduction

Thermal barrier coatings (TBCs) are an integral part of the components that operate at a high temperature, and ensures the effective lifetime of the components. The main role of the TBCs is to protect the substrate from high heat exposure and find applications in gas turbines, aeronautical machineries, and power generation equipment [1–3]. In addition to protecting the substrate, TBCs also act as a barrier towards high-temperature oxidation and corrosion resistance. The utmost importance of selecting the elements of TBCs is the similar thermal expansion coefficient. Upon experiencing numerous thermal cycles, the coatings eventually fail mechanically as a result of the generated stress between the coating and subtract [4]. To minimise this, TBCs are composed of three distinct layers on the substrate. The first layer is usually an yttria-stabilised ceramic coating, usually know as 'top coat' in the literature. This layer has low thermal conductivity and allows the growth of aluminium oxide in service and thus, provides an effective barrier against high temperature [5]. The following layer is a 'thermal growth oxide (TGO) layer', and they find their place between the 'top coat' and 'bond coat', and form via a diffusional process, to that of the elements from the bond coat [6–8]. The third layer is the 'bond coat'. The 'bond coat' in the TBCs serve two functions: (i) to avoid oxidation of the metallic substrate and (ii) to relieve the stress that is generated from the 'thermal expansion' mismatch between the substrate and 'top coat' (oxidation layer) [9].

Towards the development of the 'bond coat', a non-equimolar MCrAlY (M = Ni and/or Co) high entropy alloy (HEA) coating is promising, as stated in the literature [10]. Each of the elements in such alloy serves a specific purpose. For example, the Cr in the alloy provides resistance against high temperature corrosion. The presence of a minute amount of Y ensures solid solution hardening, in addition to improvement of the adhesion among alloy constituents. The presence of Co ensures blocking the movement of the lattice upon external loading through resistance against dislocation movements and grain boundary movements, and thus enhances the resistance against mechanical loading and creep [11]. The presence of 8–15 wt.% Al in the alloy hinders the crystal growth, which results in a stable aluminium oxide layer (α-Al_2O_3) upon high temperature exposure [12,13]. The increase in Al content in the alloy facilitates the formation of the β-(Ni,Co)Al phase and continuous α-Al_2O_3 in the MCrAlY type coatings [14]. The presence of Y in the alloy plays a secondary role by enhancing the adhesion of the alumina oxide with rest of the alloy constituents [13,15]. This improved adhesion enhances the 'cyclic oxidation resistance' of the coating [16], thanks to the presence of the reactive Y in the HEA. This also enables the use of the HEA coating in high temperature electrical insulators and sensors [17]. Traditionally, the HEAs are formed via casting into bulk form, which forms different detrimental oxide phases in the alloy. This can be avoided when the HEAs are deposited in the form of coating by 'vacuum plasma spraying' (VPS) [18,19], 'atmospheric plasma spraying' (APS) [20], or 'high-velocity oxygen fuel' (HVOF) spraying [21–23]. Among them, APS process is simple in operation and less expensive. The careful selection of spraying parameters can produce high-quality denser coating [24]. Most of the reported work on MCrAlY bond coat in the literature [16–18] focused on the temperature-dependent oxidation [25,26] and corrosion aspect [27,28] of the coating with few reports on the tribological behaviour [4,29]. However, there are no reports available on the micro-mechanical behaviours of such coatings.

As mentioned before, when exposed to the operating environment, such coating is impinged by the air, salts, debris, particles, etc., that may penetrate the porous top coat [27]. Thus, beside the investigations in regards to the oxidation and corrosion behaviours of such coating system, their micro-mechanical properties exploration is foreseen. Given the limited profile of the coatings, it is not possible to conduct traditional tensile tests on such materials deposited as coatings. This can be overcome by conducting in situ micro-pillar compression to determine the stress–strain curves under compression. This technique has been proven successful to determine the micro-mechanical properties of both bulk [30,31] and coating [32], as reported in the literature. Thus, the novelty of the present work to is investigate the stress–strain behaviour of NiCoCrAlY coating deposited by thermal spraying with the help of in situ micropillar compression, which has never been addressed before. In this research, along with the micro-mechanical properties, the structural deformation of nickel-based NiCoCrAlY HEA coating deposited on stainless steel 316 substrate was also investigated.

Thus, the aim of the present research is to explore the micro-mechanical properties of the APS-deposited NiCoCrAlY coating on stainless substrate. To achieve that, in situ compression on the micro-pillars were carried out. The associated deformation mechanisms were investigated by examining the deformed micro-pillars with an electron microscope.

2. Experimental Section

2.1. Deposition and Characterization of the Coating

In order to deposit the coating, the pre-alloyed powder of NiCoCrAlY was commercially acquired from Sulzer Metco, Winterthur, Switzerland. This specific powder composition is commercially known as Amdry 365-2. It is a Ni-rich alloy with the following composition: 47.1 wt.% Ni, 23 wt.% Co, 17 wt.% Cr, 12.5 wt.% Al, and 0.4 wt.% Y. The powder was prepared by a gas atomisation process [33] and the powder morphology is spherical, with a mean diameter of 5–38 μm as stated by the supplier [34]. Before coating deposition, the substrate was blasted by quartz sand (80–120 μm grit size), to ensure the proper adhesion of the coating with the substrate. The blasted surface was cleaned by sonication to remove any dirt/grease. The spraying gun was Diamond Jet 2700 plasma sprayer (Sulzer Metco, Switzerland), equipped with IRB 2400/16 automatic controller (ABB, Zurich, Switzerland). The optimal spraying parameters are as follows: 0.7 m^3/h hydrogen flow rate, 0.7 m^3/h argon flow rate, 40–50 g/min powder feed rate, 800 mm/s gun speed, 3 mm inter-pass spacing, 12 cm spraying distance, 38–40 bar of pressure, and with stand-off distance of 20–60 mm, and the substrate temperature was <150 °C. These parameters were selected based on the information available in the literature [4,7,13,28,29], as well as on the 'trial and error' approach to fine tune the parameters, as required by the currently employed deposition instrument.

After coating deposition, the specimen was sectioned with a water-cooled diamond saw to cut into pieces to carry out hot resin mounting and metallographic polishing. The polishing was conducted in Struers automatic polishing machine with a varying degree of diamond slurry with final polishing with colloidal silica suspension. The specimen was prepared in both planar and cross-sectional direction, and the microstructure was investigated by a field emission scanning electron microscope (Quanta 450 FE-SEM, Thermofisher Scientific, Waltham, MA, USA) and Oxford energy dispersive X-ray (EDX) analysis system (Oxford Instruments, Oxford, UK). Transmission electron microscopy (TEM) was carried out with a probe-corrected Titan TEM (Thermofisher Scientific, USA) at 200 KV. Both the micro-pillars and TEM samples were prepared by a focused ion beam (FIB)-SEM (Helios Nanolab 600, Thermofisher Scientific, USA). X-ray diffractometer (XRD) with monochromatic CuKα radiation (New D8 advance, Bruker, Karlsruhe, Germany) was used to investigate the phases in the coating.

2.2. Micro-Pillar Fabrication and Stress–Strain Calculation

Details of the micro-pillars fabrication and in situ micro-pillar compression was as follows: A focused ion beam (FIB-SEM) system (Helios Nanolab 600, FEI) was used to fabricate micro-pillars with Ø 3 μm. The micro-pillars were fabricated at the middle of 30 μm pit to avoid any contact of the indenter with the surrounding rim of the pit. Rough milling was carried out at 6.5 nA current at 30 kV, with final milling at 0.46 nA at 30 kV. A flat diamond punch of Ø 5 μm mounted on in situ nanoindentation system (PI 88, Hysitron) was used to conduct the compression tests. During compression, a 3 nm/s loading rate was selected, which corresponds to a 10^{-3} s^{-1} strain rate. The unloading rate was 50 nm/s. The whole process was recorded via videos. Deformed micro-pillars were also investigated by SEM.

Applied normal force (F) and the corresponding change in pillar length (Δl) was recorded during compression by a computer-controlled program, and subsequently used to calculate engineering stress and strain according to Equations (1) and (2):

$$\sigma = \frac{F}{A_0} \tag{1}$$

where σ is engineering stress, F is normal force, and A_0 is cross-sectional area of the pillar at 25% of its height from top. As the pillars were slightly taper (<2°), thus, most probable deformation will happen closer to the top surface [32].

$$\varepsilon_E = \frac{\Delta l}{l_0} \tag{2}$$

where ε_E is engineering strain, Δl is change in pillar length, and l_0 is initial pillar length. The details of the equations can be found in the literature [35,36]. A number of micro-pillars were fabricated on both planar and cross-section direction of the coatings. To ensure reproducibility, at least five individual experiments were conducted in each direction and representative curves were presented in the paper.

3. Results and Discussion

3.1. Coating Characterization

During the thermal spraying process, the as-received NiCoCrAlY powders experienced high temperature exposure for a fraction of a second, became molten/semi-molten, and thrust towards the substrate. The high velocity impact caused the molten/semi-molten droplets to flatten and anchor on the substrate surface. Due to this, the as-deposited coating surface is rough [37]. Thus, to unravel the microstructure on both the planar and cross-section direction, the coating was subjected to a metallographic polishing, as stated in the Experimental Section. A typical cross-sectional and planar view SEM micrograph is shown in Figure 1. The cross-sectional image of the coating (Figure 1a) shows a laminar microstructure, with a stack of splashes, which are unique of the thermal spraying-deposited material [38]. The total coating thickness was about 300 μm. However, the top section of the coating (with resin interface) is flaky in nature, which may result due to the applied force during coating deposition/hot mounting process. Withstanding that, the bottom section of the coating (with substrate interface) was dense, with an average thickness of around 145 μm. As can be seen from coating/substrate interface, the coating was well bonded with the grit-blasted substrate. In addition to that, a well-defined interface of coating/substrate was observed without significant oxide region. The planar view of the coating (Figure 2b) shows uniform microstructure composed of numerous individual splats. There were also few pores that formed on the coatings due to the trapped hot air during plasma spraying, which was below 1%, as calculated by image analysis. Some polishing debris were attached to the surface even after ultrasonic cleaning.

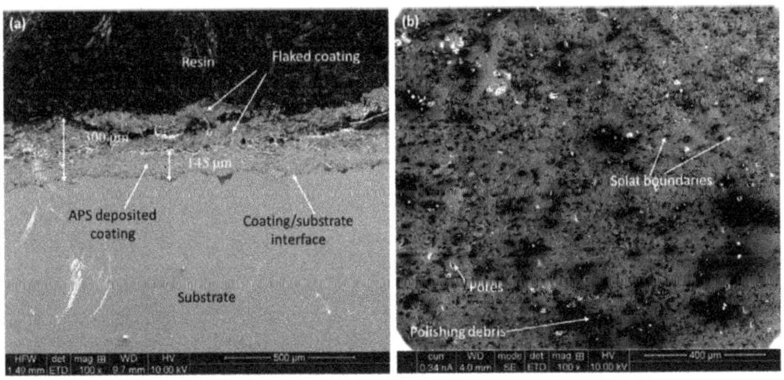

Figure 1. SEM micrographs on NiCoCrAlY coating: (**a**) cross-sectional and (**b**) planar view after metallographic polishing.

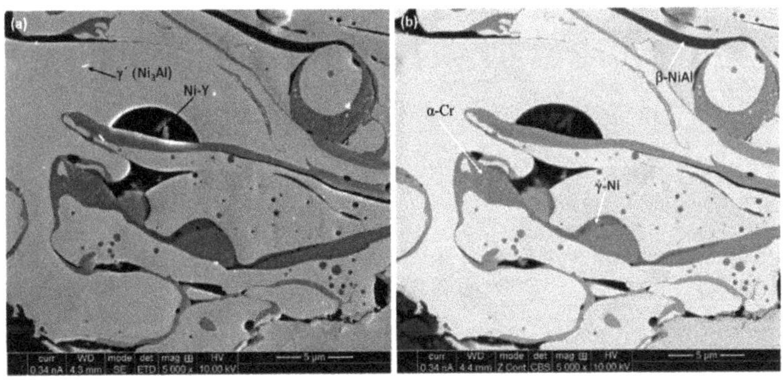

Figure 2. High-magnification SEM micrographs on the cross-section of the NiCoCrAlY coating: (**a**) secondary electron (SE) and (**b**) back-scattered electron (BSE) images after metallographic polishing.

As can be seen from Figure 1, the microstructure of the coating is complex, and resulted due to the thermal spraying process. The different elements of the coating composition own different wettability, density, and melting points, and thus, experience different amounts of heat. During thermal spraying, the feed-stock powder became molten/semi-molten due to residing in the flame, even for a fraction of a second. Thus, the homogeneous distribution of the constituent elements took place in the deposited coating, and resulted in a complex microstructure. Higher magnification secondary (SE) and back-scattered (BSE) images on the cross-section of the coating are shown in Figure 2a,b, respectively. It is evident from Figure 2 that, the multiphase NiCoCrAlY coating microstructure is made of two main phases: (i) a gamma (γ-Ni) phase matrix made of Cr and Co solid solution, and (ii) a disperse beta phase (β-NiAl) [6,32,37]. The specimen mainly consists of primary β–NiAl (BCC, gray), with a fine α–Cr (black) precipitates, β–NiAl/α–Cr eutectic, and γ–Ni (FCC, white) distributing along eutectic boundaries. The existence of such phases were confirmed by the EDX point analysis, as show in Figure 3. Moreover, it was further supported by the XRD analysis, as reported later in the section. Cuboidal Ni_3Al (γ') of several hundred nano-meters were also located along a certain orientation direction from the γ-matrix.

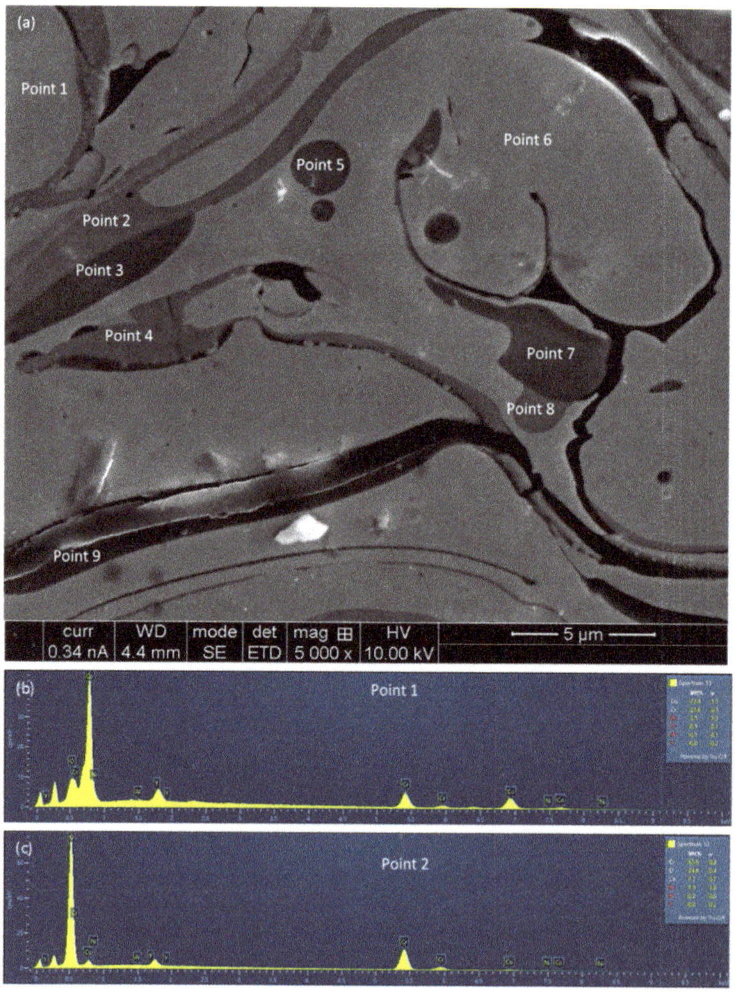

Figure 3. *Cont.*

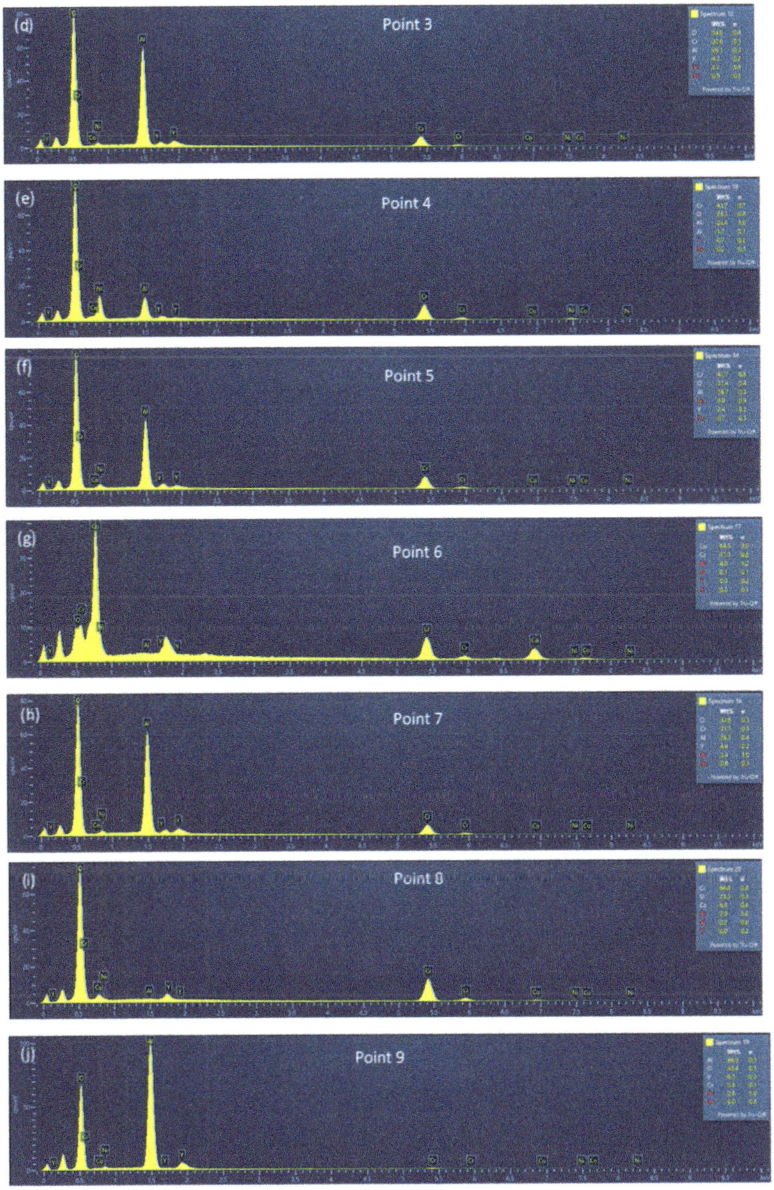

Figure 3. (**a**) High-magnification SEM micrograph on the cross-section of the NiCoCrAlY coating. The indicated points in the image are the locations of the EDX analysis. (**b–j**) EDS spectra obtained from the locations indicated in (**a**).

The elemental mapping on the cross-section of the coating is shown in Figure 4 along with the EDS spectra. Figure 4a shows the layered (overlayer) EDS images, with individual elemental images on Figure 4b–f, together with the EDS spectra in Figure 4g. As stated in the previous section, based on SEM micrograph interpretation, the coating is mainly composed of a solid solution of Ni, with a number of other phases/intermetallics in it. A peak of oxygen was identified from the EDS spectra (Figure 4g), which resulted from the oxidation of the elements, in the course of the thermal spraying. Due to that, various forms of oxides formed in the coating, which is unavoidable [3]. Among the different coating constituents, Al is prone to oxidation due to its high affinity towards oxygen to form oxides. Therefore, the amount of Al, detected by the EDS, is in the lower range of the content. In contrast to the microstructure of the bulk alloy of similar composition, the extent of secondary phase/precipitation is much restricted in the coating. This prevailed due to the rapid solidification, where time-dependent phase/precipitation was limited [39–43]. It is worthy to note that the Y content was below the detection limit of the instrument, and thus, was not quantifiable.

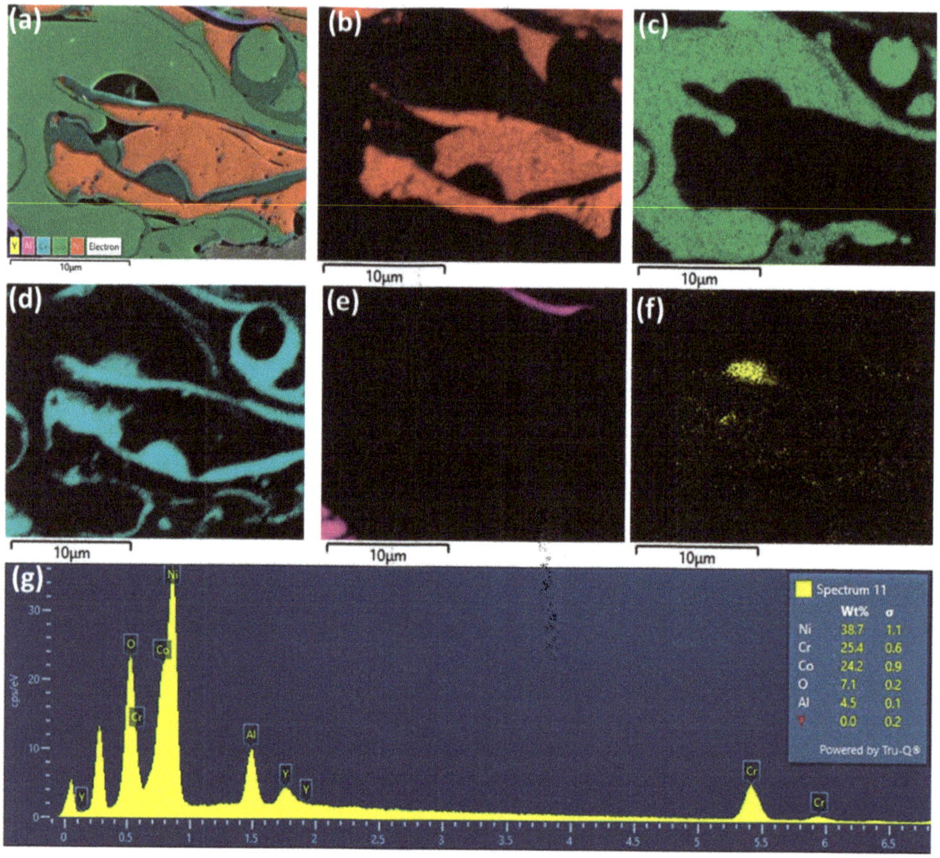

Figure 4. Elemental mapping on the cross-section of the NiCoCrAlY coating: (**a**) EDS layered image, (**b**) Ni map, (**c**) Co map, (**d**) Cr map, (**e**) Al map, (**f**) Y map, and (**g**) map sum spectrum.

Similar to that, the elemental mapping on the planar view of the coating is shown in Figure 5, with the associated EDS layered map (Figure 5a), individual elemental distribution (Figure 5b–f), and EDS spectra (Figure 5g). Similar to that appearance on the cross-section, the planar view also confirmed the presence of a solid solution, together with the splat boundaries and heterogeneous distribution of the different phases.

The XRD spectra that was obtained on the coating is shown in Figure 6. The XRD spectra confirms the existence of the different phases, which were identified by the EDS analysis. The main phases of the coating were σ-(Ni, Co, Cr), γ-Ni/γ′-Ni_3Al, β-NiAl, and various oxides (mainly Al_2O_3 and Y_2O_3).

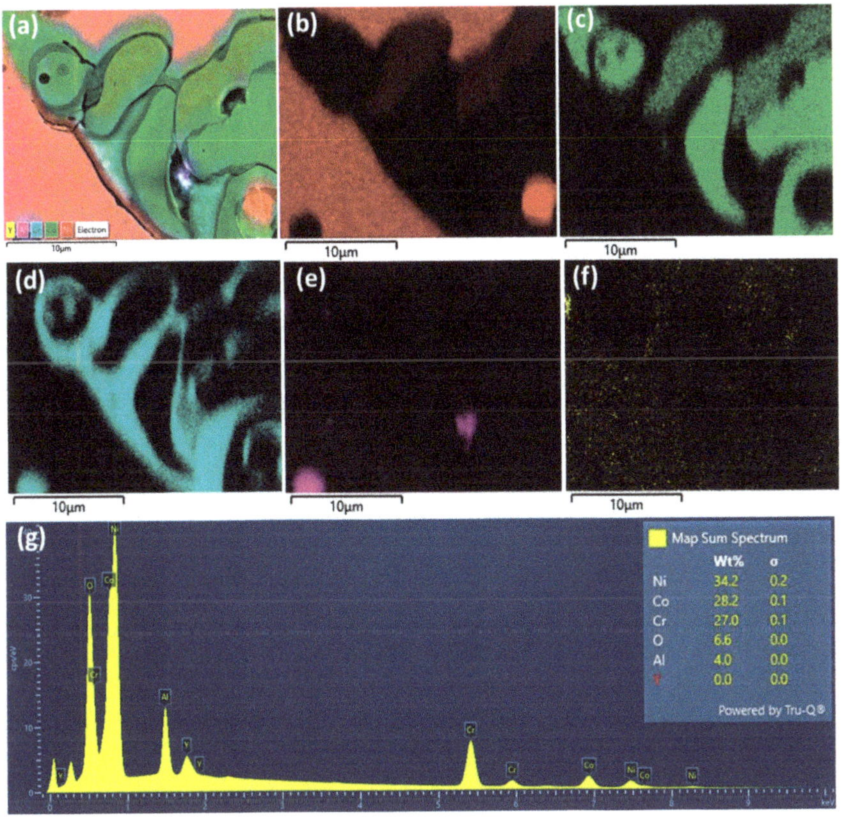

Figure 5. Elemental mapping on the surface (planar) of the NiCoCrAlY coating: (**a**) EDS layered image, (**b**) Ni map, (**c**) Co map, (**d**) Cr map, (**e**) Al map, (**f**) Y map, and (**g**) map sum spectrum.

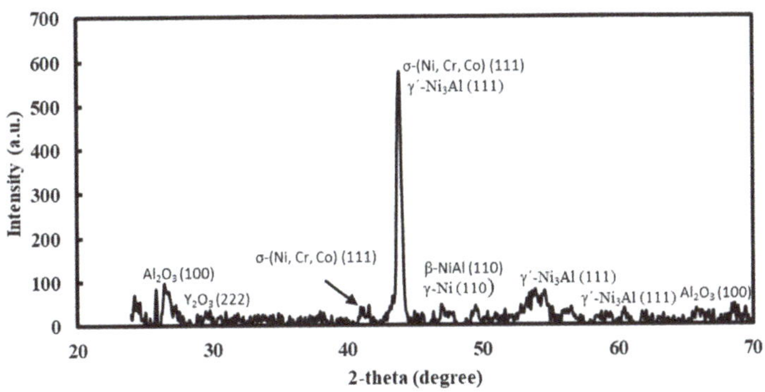

Figure 6. X-ray diffraction spectra on as-sprayed NiCoCrAlY coating.

A typical TEM micrograph of the coating in the cross-section is shown in Figure 7. A relatively low-magnification TEM bright field image (Figure 7a) exposes the lamellar-type microstructure of the thermal-sprayed coatings. As stated in the literature, exposure of the agglomerated powders at high temperatures (even for a fraction of seconds) and kinetics during thermal spraying [44–46] flattens individual molten/semi-molten droplets of powders against the substrate, producing thin layers or lamellae, often called "splats" [47].

These splats consequently stick on substrate by means of mechanical interlocking as substrate surface is at relatively low temperature than that of the splats. Localised diffusion may also take place to some extent. The constant stream of splats builds up with times, and form thick coatings, having a lamellar structure, constituted from individual thin splats". The coating thickness was built up by the accumulation of the individual splats with thickness <100 nm (Figure 7a). Higher magnification images (Figure 7b,c) show the existence of the precipitates (marked with circles), individual splats, and eutectic regions (as marked with arrows). Representative high-resolution TEM (HRTEM) image on a precipitate (Figure 7d) confirms the coherency of the precipitate with the matrix, which are metallurgically bonded with the substrate. The selected area diffraction (SAD) parent (as an insert in Figure 7d) was indexed as γ'-Ni_3Al on 101 zone axis. The diffraction confirms the polycrystalline aspect of the coating. Besides the γ'-Ni_3Al, the other nano-metric precipitation can be associated with the β-NiAl intermetallic phase (around 5–10 nm), which is common in Ni-based superalloys. Thus, it can be concluded that the coating structure was composed of two main phases, which were also supported by the findings in the literature [34].

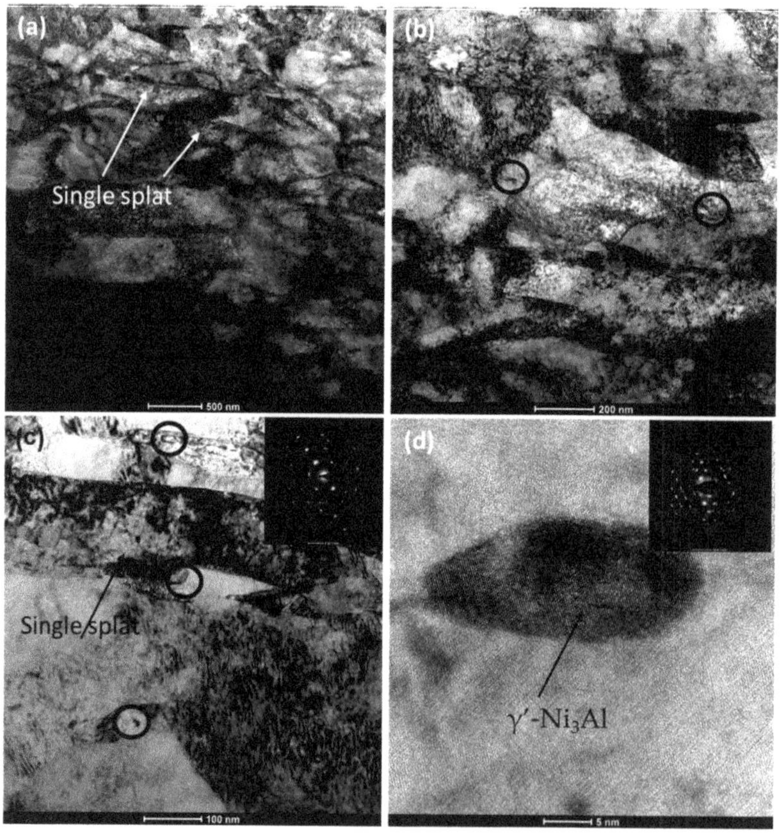

Figure 7. Representative TEM micrographs on the coating cross-section: (**a**) overall all view of the coating showing individual 'splats', (**b**,**c**) higher magnification images, and (**d**) HR-TEM image with corresponding SAD pattern of γ'-Ni_3Al on 101 zone axis, as an insert.

3.2. In Situ Micro-Pillars Compression

The micro-pillars were fabricated on both the cross-section and the surface of the coatings and disclosed in Figure 8, with higher magnification images as inserts. A 1:3

aspect ratio was maintained to avoid buckling during compression [48]. As evident from Figure 8, the pillars were somewhat tapered (<2°), which was unavoidable due to the ion beam–material interaction [49,50]. At least six individual micro-pillars were fabricated and compressed in each case to ensure data reproducibility.

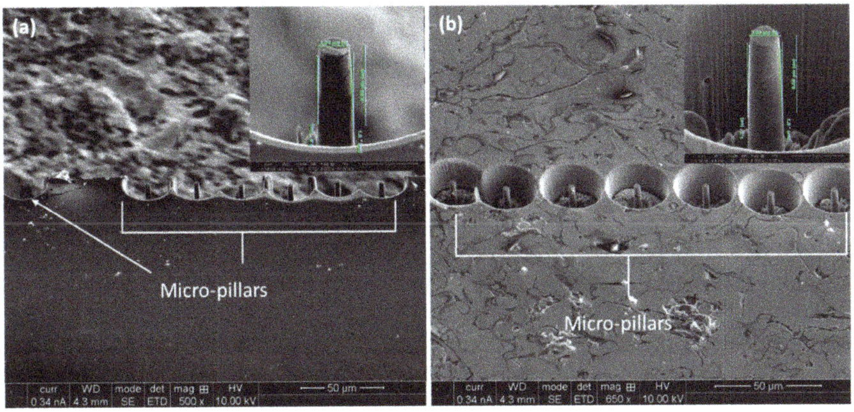

Figure 8. A series of as-fabricated micro-pillars on the (**a**) cross-section and (**b**) planar direction of the coating. Enlarged 45° view of the micro-pillars are shown as an insert.

The computer-controlled software logged the load and displacement during compression, which was then converted in to engineering stress–strain curves (Section 2.2). Figure 9 shows such engineering stress–strain curves in both planar and cross-sectional direction. Though several micro-pillars were compressed, only two representative curves in each case are reported in Figure 9, for a better view/comparison of the curves. However, for micro-mechanical properties calculation, all the graphs were considered, and an average value with standard deviation was reported.

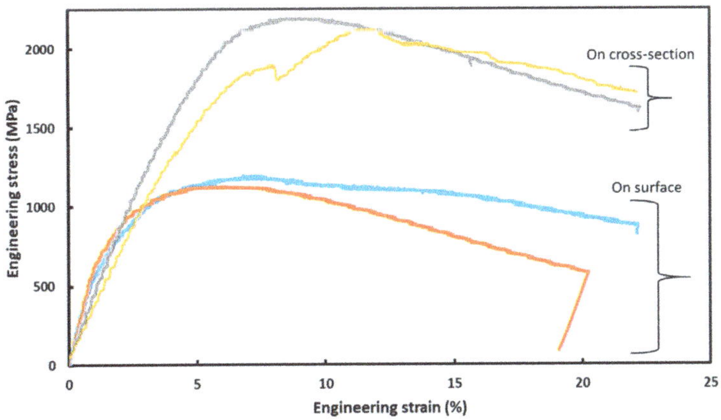

Figure 9. Average engineering stress–strain curves on the planar and cross-section direction of the NiCoCrAlY coating subjected to in situ micro-pillar compression.

As evident from Figure 9, though the curves on the cross-section and planar direction exhibit similar trends, the magnitude varies considerably. The magnitude of the curves in the cross-section is almost double that of the planar direction. At first, on the introduction of the external loading (compression), the stress increases almost linearly to that of the strain, until it reached the maximum, which is the ultimate compressive strength (UCS).

For the curves in the cross-section, the strain was about 7.5% to reach that UCS, whereas for the curves in the planar direction, it was about 5%. After that, the curves in both cases enter in the regions where the stress decreases gradually with the increase till the end of the test. The slight difference among the curves, within a given group, can attributed to the local differences in microstructure. The micro-mechanical properties of the coating as derived from the stress–strain curves are presented in Table 1. The initial linear portion of the stress–strain curves were used to calculate the elastic modulus. As can be seen from Table 1, both yield and ultimate compressive strength in the cross-section is higher than that of the planar direction (1.27 ± 0.10 MPa vs. 0.85 ± 0.09 MPa and 2.19 ± 0.10 GPa vs. 1.20 ± 0.08 GPa, respectively). However, the elastic modulus is comparable in both directions (340.24 ± 31.47 MPa vs. 331.04 ± 29.87 MPa). It was not possible to compare the presented micro-mechanical properties with any literature data, as there were no previous reports available in the literature on that, as of the authors' best knowledge.

Table 1. Mechanical properties of the NiCoCrAlY coating in the cross-section and planar direction calculated from the stress–strain curves.

NiCoCrAlY Coating	Yield Strength (σ_y), MPa	Ultimate Compressive Strength (σ_{UCS}), GPa	Elastic Modulus (E), MPa
Planar direction	0.85 ± 0.09	1.20 ± 0.08	331.04 ± 29.87
Cross-sectional direction	1.27 ± 0.10	2.19 ± 0.10	340.24 ± 31.47

3.3. Analysis of Deformed Micro-Pillars

The exact same micro-pillars, before and after the compression, were compared against each other. Figure 10 represents such comparison of three individual micro-pillars in the cross-section direction. As can be seen from the first row of the images (Figure 10a–c), the micro-pillars contain certain features, such as phase boundaries of different size and shape, before the compression. There was no visible deformation in the initial stage, where the stress rises proportionally to the strain (<7.5% strain). As the compression continued, the pre-existing phase boundaries acted as the 'weakest link' and served as a stress raise/crack initiator point. As a 'weakest link' in the structure, cracks preferentially took place in these regions in the form of slip planes.

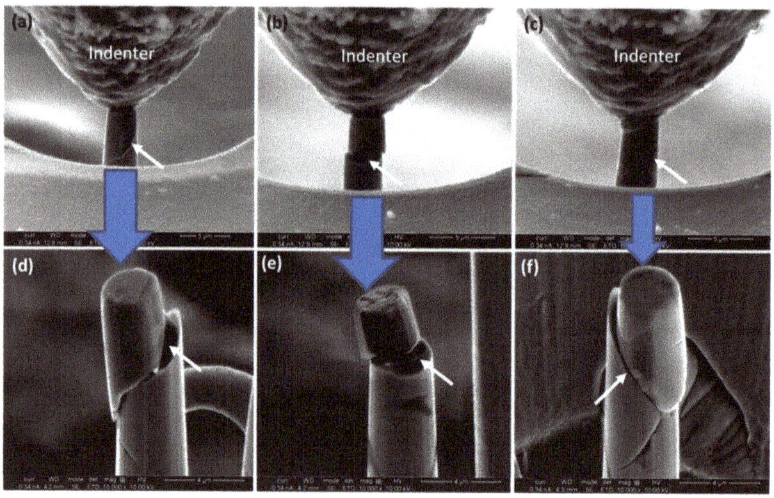

Figure 10. Deformed micro-pillars (45° tilted SEM view) after compression on cross-sectional direction. State of the micropillars at the beginning of compression (**a**–**c**) and at the end of compression (**d**–**f**). The white arrows indicate the fracture lines.

Similar scenarios were also observed in the micro-pillars in planar direction, as shown in Figure 11. As the deformation continues, the micro-pillars can no longer lodge the external loading. As mentioned previously, it is interesting to note that, the ultimate separation (through fracture) took place along the pre-existing phase boundaries in both cases. In the case of a specific micro-pillar (Figure 11b,e), there was no pre-existing secondary phase (Figure 11b). In this particular case, there was no 'through' fracture, and the micro-pillar just got squeezed. Thus, it can be concluded that the presence of the pre-existing secondary phases dominates the fracture path of the deformed micro-pillars.

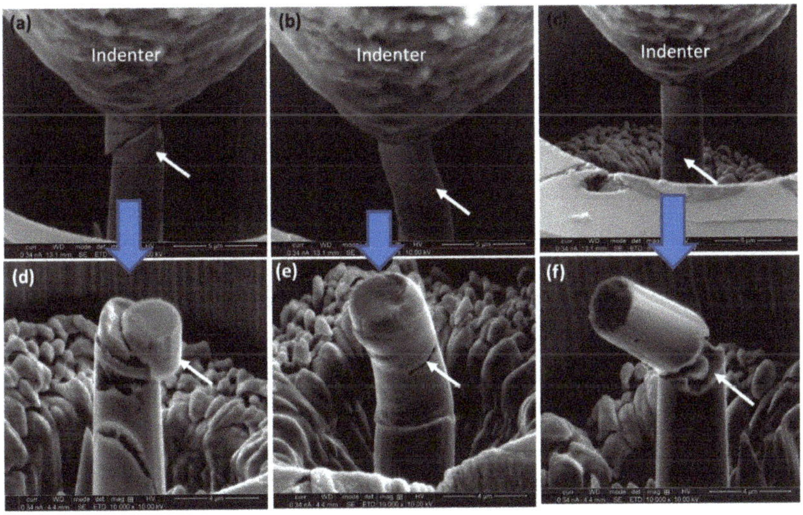

Figure 11. Deformed micro-pillars (45° tilted SEM view) after compression on planar direction. State of the micropillars at the beginning of compression (**a–c**) and at the end of compression (**d–f**). The white arrows indicate the fracture lines.

3.4. Deformation Mechanisms

It is important to note that formation and presence of the secondary/ intermetallic phases in the present investigated coating provides a strengthening effect, both at room and at high temperatures [34]. This resembled that of "metal matrix composites [30], where reinforcing particles provide strength, i.e., load bearing capacity, and metal matrix provide toughness, i.e., accommodation of external loading'. Presence of such reinforced particles restrict dislocation movement, and thus increases the strength of the composite, due to Orowan strengthen mechanisms". The deformation aspect of the coating can be discussed in view of two separate aspects: (i) the contribution of the secondary phase/intermetallic particles, and (ii) loading direction with respect to splat orientation. As discussed in Section 3.1, the surface (planar) and cross-sectional view of the coating microstructure show two different aspects. During compression in the planar direction, the flat punch (indenter) is perpendicular towards the stacked-up splats. As the loading continued, individual splats experienced stress in both radial and vertical direction, and squeezed till cracks took place both through, and along the splat boundaries. On the micro-pillar surface, this is evident as wrinkles (Figure 11e). On the other hand, loading direction is parallel to stacked-up splats on the cross-sectional direction. Thus, formation and propagation of cracks are favourable. These scenarios get more complicated in the presence of the pre-existing phase boundaries in the micro-pillars, which act as the 'weakest link' in the structure, as established before. Relative movement took place along the boundaries of those phases, with the rest of the material of the structure. Once the stress level reaches the 'critically resolved shear stress', the formation of the slip/shear planes occur. The phase boundaries are metallurgically bonded with the matrix, whereas the splash boundaries are mostly mechanically interlocked. Thus, a relative competition took place regarding which mode of crack propagation will be prevailing.

The above-mentioned mechanism of deformation is supported by the evidence on the stress–strain curves (Figure 9), together with the post-deformation examination of the micro-pillars (Figures 10 and 11). In summary, the presence of both the phase and splat boundaries in the coating structure, and their orientation with respect to loading direction dominate the occurrence and propagation of the cracks. This also heavily contributes on the micro-mechanical properties of the coating. The 'wrinkles (steps)' are more pronounced on the deformed pillar surface in the absence of the secondary phases. This means that coating shows relatively lower resistance against deformation.

4. Conclusions

The NiCoCrAlY HEA coating was successfully deposited on the stainless steel subtract. The microstructural evolution of the coating was invested by electron microscopy, and the micro-mechanical properties of the coating was investigated by in situ micro-pillar compression, both in the planar and cross-sectional. Besides that, the deformation aspect of the coating under compression was also explored. Based on the experimental proofs, the following conclusion could be drawn as an outcome of this present study:

1. APS-deposited NiCoCrAlY HEA coating attained metallurgical bonding with the substrate, and the as-deposited coating thickness was about 300 μm, with limited porosity (<1%). The coating structure revealed representative lamellar-like structure that are characteristics of thermal-deposited coating, which consisted of the Ni (γ-Ni) matrix solid solution and secondary phase, like β–NiAl, together with the precipitates fine α–Cr and γ'–Ni_3Al intermetallics.
2. The strength and modulus of elasticity of the coating was higher in the cross-sectional direction (1.27 ± 0.10 MPa of yield strength and 2.19 ± 0.10 GPa of compressive strength), than that of the planar direction (0.85 ± 0.09 MPa of yield strength and 1.20 ± 0.08 GPa of compressive strength). The striking difference was due to the presence of the secondary/intermetallic phases in the micro-pillar, together with that of the loading direction with respect to the splat boundaries orientation.
3. The deformation of the coating took place as a result of the wrinkle and slip/shear bands formation once the mechanical loading exceeded the strength of the material.

Author Contributions: A.K.B.: conceptualization, writing, experiment, and supervision; N.R.: resource and writing—review and editing; C.P.: formal analysis, validation, and writing—review and editing; A.P.: writing—review and editing, formal analysis, and validation. All authors have read and agreed to the published version of the manuscript.

Funding: This research received no external funding.

Institutional Review Board Statement: Not applicable.

Informed Consent Statement: Not applicable.

Data Availability Statement: The raw/processed data used to produce the results will be made available by the corresponding author upon reasonable request.

Conflicts of Interest: The authors declare that they have no known competing financial interest or personal relationship whatsoever that could have influenced the work reported in this paper.

References

1. Padture, N.P.; Gell, M.; Jordan, E.H. Thermal barrier coatings for gas-turbine engine applications. *Science* **2002**, *296*, 280–284. [CrossRef] [PubMed]
2. Beele, W.; Marijnissen, G.; Van Lieshout, A. The evolution of thermal barrier coatings—Status and upcoming solutions for today's key issues. *Surf. Coat. Technol.* **1999**, *120*, 61–67. [CrossRef]
3. Cao, X.; Vassen, R.; Stöver, D. Ceramic materials for thermal barrier coatings. *J. Eur. Ceram. Soc.* **2004**, *24*, 1–10. [CrossRef]

4. Hao, E.; An, Y.; Zhao, X.; Zhou, H.; Chen, J. NiCoCrAlYTa coatings on nickel-base superalloy substrate: Deposition by high velocity oxy-fuel spraying as well as investigation of mechanical properties and wear resistance in relation to heat-treatment duration. *Appl. Surf. Sci.* **2018**, *462*, 194–206. [CrossRef]
5. Smialek, J.L.; Miller, R.A. Revisiting the birth of 7YSZ thermal barrier coatings: Stephan Stecura. *Coatings* **2018**, *8*, 255. [CrossRef]
6. Haynes, J. *Elevated Temperature Coatings: Science and Technology*; TMS: Warrendale, PA, USA, 1999; pp. 185–196.
7. Kim, J.; Pyeon, J.; Kim, B.-G.; Khadaa, T.; Choi, H.; Zhe, L.; Dube, T.; Zhang, J.; Yang, B.-i.; Jung, Y.-g. Oxidation Behavior of NiCoCrAlY Coatings Deposited by Vacuum Plasma Spraying and High-Velocity Oxygen Fuel Processes. *Coatings* **2023**, *13*, 319. [CrossRef]
8. Schulz, U.; Leyens, C.; Fritscher, K.; Peters, M.; Saruhan-Brings, B.; Lavigne, O.; Dorvaux, J.-M.; Poulain, M.; Mévrel, R.; Caliez, M. Some recent trends in research and technology of advanced thermal barrier coatings. *Aerosp. Sci. Technol.* **2003**, *7*, 73–80. [CrossRef]
9. Wang, Y.; Zhang, C.; Li, J.; Ding, G.; Duan, L. Fabrication and characterization of ITO thin film resistance temperature detector. *Vacuum* **2017**, *140*, 121–125. [CrossRef]
10. Bates, B.; Witman, J.; Zhang, Y. Electrolytic co-deposition of Ni-CrAlY composite coatings using different deposition configurations. *Mater. Manuf. Process.* **2016**, *31*, 1232–1237. [CrossRef]
11. Tancret, F.; Bhadeshia, H.; MacKay, D. Design of a creep resistant nickel base superalloy for power plant applications: Part 1-Mechanical properties modelling. *Mater. Sci. Technol.* **2003**, *19*, 283–290. [CrossRef]
12. Marginean, G.; Utu, D. Cyclic oxidation behaviour of different treated CoNiCrAlY coatings. *Appl. Surf. Sci.* **2012**, *258*, 8307–8311. [CrossRef]
13. Nijdam, T.; Kwakernaak, C.; Sloof, W. The effects of alloy microstructure refinement on the short-term thermal oxidation of NiCoCrAlY alloys. *Metall. Mater. Trans. A* **2006**, *37*, 683–693. [CrossRef]
14. Kim, D.; Shin, I.; Koo, J.; Kim, S.; Seo, D.; Kim, J.; Seok, C. Quantitative analysis on the depletion rate of β-NiAl phases in MCrAlY coating. *J. Mech. Sci. Technol.* **2014**, *28*, 513–519. [CrossRef]
15. Toscano, J.; Vaβen, R.; Gil, A.; Subanovic, M.; Naumenko, D.; Singheiser, L.; Quadakkers, W. Parameters affecting TGO growth and adherence on MCrAlY-bond coats for TBC's. *Surf. Coat. Technol.* **2006**, *201*, 3906–3910. [CrossRef]
16. Wang, B.; Gong, J.; Sun, C.; Huang, R.; Wen, L. The behavior of MCrAlY coatings on Ni3Al-base superalloy. *Mater. Sci. Eng. A* **2003**, *357*, 39–44. [CrossRef]
17. Hou, X.; Zhang, C.; Wang, F.; Ding, G. Fabrication and characterization of NiCoCrAlY coating deposited on nickel-based superalloy substrates. *Vacuum* **2018**, *155*, 55–59. [CrossRef]
18. Gudmundsson, B.; Jacobson, B.; Gruner, H. The influence of substrate temperature on the microstructure and hardness of vacuum-plasma-sprayed Co–Ni–Cr–Al–Si–Zr–Y and Co–Ni–Cr–Al–Y alloys. *Mater. Sci. Eng. A* **1989**, *108*, 105–115. [CrossRef]
19. Noguchi, K.; Nishida, M.; Chiba, A. Transmission electron microscopy of low pressure plasma sprayed CoNiCrAlY coating. *Scr. Mater.* **1996**, *35*, 1359–1364. [CrossRef]
20. Basak, A.; Matteazzi, P.; Vardavoulias, M.; Celis, J.-P. Corrosion–wear behaviour of thermal sprayed nanostructured FeCu/WC–Co coatings. *Wear* **2006**, *261*, 1042–1050. [CrossRef]
21. Basak, A.K.; Celis, J.-P.; Vardavoulias, M.; Matteazzi, P. Effect of nanostructuring and Al alloying on friction and wear behaviour of thermal sprayed WC–Co coatings. *Surf. Coat. Technol.* **2012**, *206*, 3508–3516. [CrossRef]
22. Hou, G.; An, Y.; Zhao, X.; Zhou, H.; Chen, J. Effect of alumina dispersion on oxidation behavior as well as friction and wear behavior of HVOF-sprayed CoCrAlYTaCSi coating at elevated temperature up to 1000 C. *Acta Mater.* **2015**, *95*, 164–175. [CrossRef]
23. Zhang, L.-w.; Lu, L.; Wang, L.; Ning, X.-j.; Wang, Q.-s.; Wang, R.-x. Microstructural characteristics and oxidation behavior of low-pressure cold-sprayed CoNiCrAlY coatings. *J. Therm. Spray Technol.* **2017**, *26*, 1565–1572. [CrossRef]
24. Brandl, W.; Toma, D.; Krüger, J.; Grabke, H.; Matthäus, G. The oxidation behaviour of HVOF thermal-sprayed MCrAlY coatings. *Surf. Coat. Technol.* **1997**, *94*, 21–26. [CrossRef]
25. Tang, F.; Ajdelsztajn, L.; Schoenung, J.M. Characterization of oxide scales formed on HVOF NiCrAlY coatings with various oxygen contents introduced during thermal spraying. *Scr. Mater.* **2004**, *51*, 25–29. [CrossRef]
26. Lee, C.; Kim, H.; Choi, H.; Ahn, H. Phase transformation and bond coat oxidation behavior of plasma-sprayed zirconia thermal barrier coating. *Surf. Coat. Technol.* **2000**, *124*, 1–12. [CrossRef]
27. Kawakita, J.; Kuroda, S.; Fukushima, T.; Kodama, T. Development of dense corrosion resistant coatings by an improved HVOF spraying process. *Sci. Technol. Adv. Mater.* **2003**, *4*, 281–289. [CrossRef]
28. Radhika, N.; Noble, N.; Adediran, A.A. Electrochemical and hot corrosion behaviour of annealed AlCoCrFeNi HEA coating over steel. *Sci. Rep.* **2024**, *14*, 5652. [CrossRef] [PubMed]
29. Noble, N.; Radhika, N.; Sathishkumar, M.; Basak, A. Slurry erosion behaviour of AlCoCrFeNi high entropy alloy coating prepared by atmospheric plasma spraying. *Trans. IMF* **2024**, 1–8. [CrossRef]
30. Basak, A.; Pramanik, A.; Prakash, C. Deformation and strengthening of SiC reinforced Al-MMCs during in-situ micro-pillar compression. *Mater. Sci. Eng. A* **2019**, *763*, 138141. [CrossRef]
31. Tabbakh, T.; Alshihri, S.; Basak, A.; Kurdi, A. Strength of a 3D printed Al 7068 alloy under micro-pillar compression. *Met. Mater. Int.* **2022**, *28*, 2706–2718. [CrossRef]
32. Kurdi, A.; Basak, A. Deformation of electrodeposited gradient Co/Sn multilayered coatings under micro-pillar compression. *Eng. Fract. Mech.* **2018**, *204*, 138–146. [CrossRef]

33. Catto, F.; Gabriel, A.; Bolfarini, C.; Kiminami, C.S.; Afonso, C.R.M. Rapid solidification and laser cladding of gas atomized Ni-Nb-Sn bulk metallic glass. In *Materials Science Forum*; Trans Tech Publications Ltd.: Stafa-Zurich, Switzerland, 2017; pp. 311–316.
34. Falcón, J.C.P.; Echeverria, A.; Afonso, C.R.; Carrullo, J.C.Z.; Borrás, V.A. Microstructure assessment at high temperature in NiCoCrAlY overlay coating obtained by laser metal deposition. *J. Mater. Res. Technol.* **2019**, *8*, 1761–1772. [CrossRef]
35. Kiener, D.; Motz, C.; Dehm, G. Micro-compression testing: A critical discussion of experimental constraints. *Mater. Sci. Eng. A* **2009**, *505*, 79–87. [CrossRef]
36. Girault, B.; Schneider, A.S.; Frick, C.P.; Arzt, E. Strength effects in micropillars of a dispersion strengthened superalloy. *Adv. Eng. Mater.* **2010**, *12*, 385–388. [CrossRef]
37. Tang, J.; Bai, Y.; Zhang, J.; Liu, K.; Liu, X.; Zhang, P.; Wang, Y.; Zhang, L.; Liang, G.; Gao, Y. Microstructural design and oxidation resistance of CoNiCrAlY alloy coatings in thermal barrier coating system. *J. Alloys Compd.* **2016**, *688*, 729–741. [CrossRef]
38. Liang, J.; Wei, H.; Hou, G.; Zheng, Q.; Sun, X.; Guan, H.; Hu, Z. Thermal stability of phases in a NiCoCrAlY coating alloy. *J. Mater. Res.* **2008**, *23*, 2264–2274. [CrossRef]
39. Yuan, K.; Peng, R.L.; Li, X.-H.; Johansson, S. Hot corrosion behavior of HVOF-sprayed CoNiCrAlYSi coatings in a sulphate environment. *Vacuum* **2015**, *122*, 47–53. [CrossRef]
40. Huntz, A. Influence of active elements on the oxidation mechanism of M–Cr–Al alloys. *Mater. Sci. Eng.* **1987**, *87*, 251–260. [CrossRef]
41. Stott, F.; Wood, G. Growth and adhesion of oxide scales on Al2O3-forming alloys and coatings. *Mater. Sci. Eng.* **1987**, *87*, 267–274. [CrossRef]
42. Smeggil, J. Some comments on the role of yttrium in protective oxide scale adherence. *Mater. Sci. Eng.* **1987**, *87*, 261–265. [CrossRef]
43. Volenik, K.; Novak, V.; Dubský, J.; Chraska, P.; Neufuss, K. Properties of alloy steel coatings oxidized during plasma spraying. *Mater. Sci. Eng. A* **1997**, *234*, 493–496. [CrossRef]
44. Buschow, K. *Encyclopedia of Materials: Science and Technology*; Elsevier: Amsterdam, The Netehrlands, 2001.
45. Basak, A.; Celis, J.-P.; Ponthiaux, P.; Wenger, F.; Vardavoulias, M.; Matteazzi, P. Effect of nanostructuring and Al alloying on corrosion behaviour of thermal sprayed WC–Co coatings. *Mater. Sci. Eng. A* **2012**, *558*, 377–385. [CrossRef]
46. Basak, A.; Achanta, S.; De Bonte, M.; Celis, J.-P.; Vardavoulias, M.; Matteazzi, P. Effect of Al and Cr addition on tribological behaviour of HVOF and APS nanostructured WC–Co coatings. *Trans. IMF* **2007**, *85*, 310–315. [CrossRef]
47. Basak, A.; Achanta, S.; Celis, J.-P.; Vardavoulias, M.; Matteazzi, P. Structure and mechanical properties of plasma sprayed nanostructured alumina and FeCuAl–alumina cermet coatings. *Surf. Coat. Technol.* **2008**, *202*, 2368–2373. [CrossRef]
48. Misra, A.; Hirth, J.; Hoagland, R. Length-scale-dependent deformation mechanisms in incoherent metallic multilayered composites. *Acta Mater.* **2005**, *53*, 4817–4824. [CrossRef]
49. Wang, J.; Stanford, N. Investigation of precipitate hardening of slip and twinning in Mg5% Zn by micropillar compression. *Acta Mater.* **2015**, *100*, 53–63. [CrossRef]
50. Zhang, J.; Liu, G.; Lei, S.; Niu, J.; Sun, J. Transition from homogeneous-like to shear-band deformation in nanolayered crystalline Cu/amorphous Cu–Zr micropillars: Intrinsic vs. extrinsic size effect. *Acta Mater.* **2012**, *60*, 7183–7196. [CrossRef]

designs

MDPI

Article

Computational Investigation of the Fluidic Properties of Triply Periodic Minimal Surface (TPMS) Structures in Tissue Engineering

Muhammad Noman Shahid [1], Muhammad Usman Shahid [1], Shummaila Rasheed [1], Muhammad Irfan [1] and Muhannad Ahmed Obeidi [2,*]

1 Department of Mechanical Engineering, Capital University of Science and Technology, Islamabad 45750, Pakistan; nomanshahid774@gmail.com (M.N.S.); usmanshahid866@gmail.com (M.U.S.); shummaila@cust.edu.pk (S.R.); m.irfan@cust.edu.pk (M.I.)

2 School of Mechanical and Manufacturing Engineering, Dublin City University, D09 V209 Dublin, Ireland

* Correspondence: obeidimuhannad@gmail.com or muhannad.ahmedobeidi@dcu.ie

Abstract: Tissue engineering, a rapidly advancing field in medicine, has made significant strides with the development of artificial tissue substitutes to meet the growing need for organ transplants. Three-dimensional (3D) porous scaffolds are widely utilized in tissue engineering, especially in orthopedic surgery. This study investigated the fluidic properties of diamond and gyroid structures with varying porosity levels (50–80%) using Computational Fluid Dynamics (CFD) analysis. The pressure and velocity distributions were analyzed, and it was observed that the pressure decreased gradually, whereas the velocity increased in the central area of the surface structures. Specifically, the pressure drop ranged from 2.079 to 0.984 Pa for the diamond structure and from 1.669 to 0.943 Pa for the gyroid structure as the porosity increased from 50% to 80%. It was also found that the permeability increased as the porosity level increased, with values ranging from 2.424×10^{-9} to 5.122×10^{-9} m^2 for the diamond structure and from 2.966×10^{-9} to 5.344×10^{-9} m^2 for the gyroid structure. The wall shear stress (WSS) was also analyzed, showing a consistent decrease with increased porosity for both types of structures, with WSS values ranging from 9.903×10^{-2} to 9.840×10^{-1} Pa for the diamond structure and from 1.150×10^{-1} to 7.717×10^{-2} Pa for the gyroid structure. Overall, this study provides insights into the fluidic properties of diamond and gyroid structures, which can be useful in various applications such as tissue engineering.

Keywords: tissue engineering; computational fluid dynamics; triply periodic minimal surfaces; gyroid; diamond; porosity; wall shear stress; permeability

1. Introduction

In recent years, tissue engineering has seen substantial advancements, driven by the creation of novel artificial tissue replacements designed to address the increasing demand for organ transplants. Particularly in orthopedic surgery, three-dimensional (3D) porous scaffolds with diverse structural designs are frequently employed in tissue engineering. Advancements in manufacturing technologies, including additive manufacturing, have enabled the production of 3D scaffolds with a variety of materials and structures. Nevertheless, achieving the ideal scaffold design remains challenging, as the design parameters significantly influence the scaffold's mechanical properties and biocompatibility [1,2]. For example, while a high porosity is favorable for biological compatibility, it may compromise the mechanical integrity of the scaffold. It is also necessary to properly design the mass flow characteristics of porous structures, such as permeability, to permit the passage of nutrients and oxygen into cells inside the structure [3,4].

In tissue engineering, it is a common practice to predict the permeability of scaffolds using computer models. Studies have shown that porous structures can serve as scaffolds to facilitate osteogenic differentiation, adhesion, proliferation, host integration, cell attachment, and cell viability. To facilitate nutrient diffusion, cell infiltration, and the removal of metabolic waste products, scaffolds must exhibit certain characteristics. They should have interconnected pores and a high level of porosity [5]. It is difficult to develop a porous structure that meets the requirements of both mechanical and biological functions. However, the geometry and curvature of the surfaces where cells reside are recognized as crucial factors in evaluating the rate of tissue production. Recent theories suggest that, unlike flat and convex surfaces, high-curvature surfaces and concave surfaces, respectively, promote and enhance tissue regeneration [6]. Additionally, surface roughness is an important quality measure as it supports cell differentiation and growth, while also potentially localizing stress. These requirements can be addressed by designing scaffolds based on Triply Periodic Minimal Surfaces (TPMS) [3,7].

A TPMS, or the surface of the minimal area, possesses a boundary delineated by a closed curve [8]. TPMS scaffolds are constructed on this surface, with each unit cell created by adding thickness to the surface and subsequently combining to form a final scaffold that exhibits cubic symmetry and comprises periodically repeated interconnected pores. Therefore, a TPMS is more desirable for bone regeneration [9], and has characteristics similar to the curvature of the trabecular bone [10]. Several natural geometries such as cubosomes, cell membranes, and plant prolamellar bodies can be characterized by a TPMS [7]. Previous research has investigated how various pore sizes impact the bond strength between an implant and bone. Their findings indicated that pore sizes between 300 μm and 600 μm were optimal in facilitating continuous bone growth to enhance osseointegration. The largest pore size demonstrated the quickest fluid movement, indicating that increasing pore size can enhance the rate of cell migration [11].

Evaluating the characteristics related to biocompatibility, such as the permeability and effect of wall shear stress, is of the utmost importance because of their impact on the biological activity of cells within the supportive structures [12]. Numerous investigations have been undertaken to measure the permeability of scaffolds and analyze the factors that impact this permeability. A scaffold that allows substances such as nutrients and gasses to pass through its pores adequately while also enabling the effective removal of waste is considered permeable [13]. Pires et al. found that calculating permeability is crucial for optimizing cell cultivation and determining treatment duration, as it directly influences cell growth and proliferation within the supportive structures [14]. Ma et al. revealed that, by modifying the design parameters, it is possible to adjust the permeability and mechanical characteristics of the Gyroid structure to mimic those found in trabecular bones within the human body. Furthermore, they observed that permeability had a significant impact on initial cell growth [15]. When designing a scaffold, permeability is a vital consideration, as it significantly influences the transport of oxygen, nutrients, and the removal of metabolic waste. Geometric factors, including pore size, surface thickness, and structure type, also affect permeability [5,16,17]. Researchers have investigated the concept of developing the microstructure of scaffolds through topology optimization. The findings demonstrated a direct relationship between computational and experimental permeability, underscoring the significance of permeability as a crucial design factor. The study illustrated that the regeneration of tissue is influenced by factors such as porosity and permeability, highlighting their potential for furthering our understanding of the impact of scaffold architecture on biological processes [18]. Aqila et al. [19] conducted a study to examine how the size of pores influences the mechanical behavior and fluid permeability of bone scaffolds. Their findings indicated that larger pore sizes were associated with improved permeability and reduced wall shear stress across different flow rates, facilitating a more effective transportation of fluids and nutrients. Ntousi et al. [20] used a computational fluid dynamic (CFD) analysis to analyze two scaffold structures with a pore size of 500 μm. The results revealed that the designed scaffolds exhibited enhanced permeability and fluid flow, making them advantageous for bone tissue engineering applications. Pires et al. [14] evaluate the permeability of various TPMS scaffold geometries through a combination of experimental pressure drop tests and CFD simulations. The results suggested that Gyroid structures were the most optimal choice for applications in bone tissue engineering, as they consistently demonstrated fluid permeation, closely followed by Schwarz D. Conversely, Schwarz P configurations displayed uniform flow patterns and notable fluctuations in permeability based on porosity levels.

Another critical factor in studying the fluidic response of scaffolds is Wall Shear Stress (WSS). WSS has been recognized for its impact on cell growth [8,21]. Fluid-induced WSS provides a mechanical stimulus that can be sensed by cells, leading to enhanced cell differentiation and proliferation [22]. Ali et al. [23] conducted a comparative analysis using CFD simulations to assess the permeability and WSS of TPMS and lattice-based scaffolds. The analysis revealed how internal structures affect the behavior of fluids. Lesman et al. [24] conducted CFD simulations to demonstrate that when cells grow within 3D scaffolds during perfusion cell culturing, the WSS increases as a result of contraction in the channels of the scaffolds. Marin et al. [25] conducted a CFD analysis to investigate how the variation in structure among scaffolds affects the distribution and magnitude of WSS. The findings of their research revealed that the presence of geometric inconsistencies led to significant variations in velocity and WSS across the different samples. A CFD study was conducted in scaffolds with square pores of varying sizes, with a constant porosity of 63% to evaluate the permeability and WSS. The research revealed that s higher surface roughness led to a decline in both permeability and WSS. It is noteworthy that there was a notable decrease in WSS observed when comparing smooth and rough scaffolds [3]. Zhao et al. [26] conducted a study that analyzed the impact of the geometric characteristics of scaffolds using a fluid–structure interaction model. The results showed that pore size significantly influences mechanical stimulation, and simultaneous fluid perfusion and mechanical load increase WSS. In addition, Lee et al. [27] utilized the Allen–Cahn equation to observe and analyze the dynamic growth behaviors of cells within various scaffold structures. Although various authors have explored the impacts of the geometric aspects of TPMS structures on mechanical performance [22,28,29], studying the fluid behavior of different TPMS structures remains a promising area of research, particularly in applications related to bone implants.

The primary objective of this research is to create TPMS structures utilizing nTopology 4.24.3 software and subsequently optimize them. Following this, the optimized scaffolds, namely Diamond and Gyroid, which have different levels of porosity, are evaluated to determine their fluidic flow characteristics. CFD analysis is employed to examine physical characteristics including pressure differentials, permeability, and WSS. These parameters provide a basis for investigating the biological behavior of these structures. Our investigation focuses on evaluating the flow characteristics of these structures. The goal is to assess their potential as bone scaffolds and contribute to the advancement of optimized bone tissue engineering applications.

2. Materials and Methods

Initially, a scaffold was developed leading to the creation of a three-dimensional model. Following these procedural steps, a CFD analysis was undertaken on the scaffolds, investigating the fluidic behavior of blood under varying porosity levels.

2.1. Scaffold Design

The design of scaffolds is critical in CFD studies because of the behavior of fluidic flow, which is dependent on lattice structure. Two architectural scaffolds, namely diamond and gyroid, were selected and developed utilizing the nTopology software having different porosities ranging from 50% to 80% with an increment of 10% (Figure 1). The scaffolds were subsequently exported to ANSYS SpaceClaim, where they were converted from STL to solids. The different porosities in the diamond and gyroid structures are achieved by altering the wall thickness shown in Table 1 Thicker walls reduce the void space, resulting in lower porosity, while thinner walls increase the void space, leading to higher porosity.

The cells display symmetry along each of the three orthogonal axes (x, y, and z), making them isotropic in their geometric properties.

CFD analysis was conducted using the ANSYS Fluent 2021 R2 module. To streamline computational time, each model was maintained at a size of 2 mm × 2 mm × 2 mm, taking into account the geometric complexity of the model. The 2 mm × 2 mm × 2 mm size is optimal for computational models due to its balance between efficiency and capturing the geometric intricacies of bone scaffolds [6]. While not reflecting all clinical scenarios, it represents a subset for analyzing fundamental fluid behavior within computational constraints. For larger scaffold structures, we emphasize the importance of cautious evaluation and experimental validation. Our study aims to provide critical insights on fundamental fluid dynamics principles, which are essential for a wide range of clinical scaffold sizes. In Figure 1, the subscript indicates the porosity percentage, like D_{50} which represents 50% porosity.

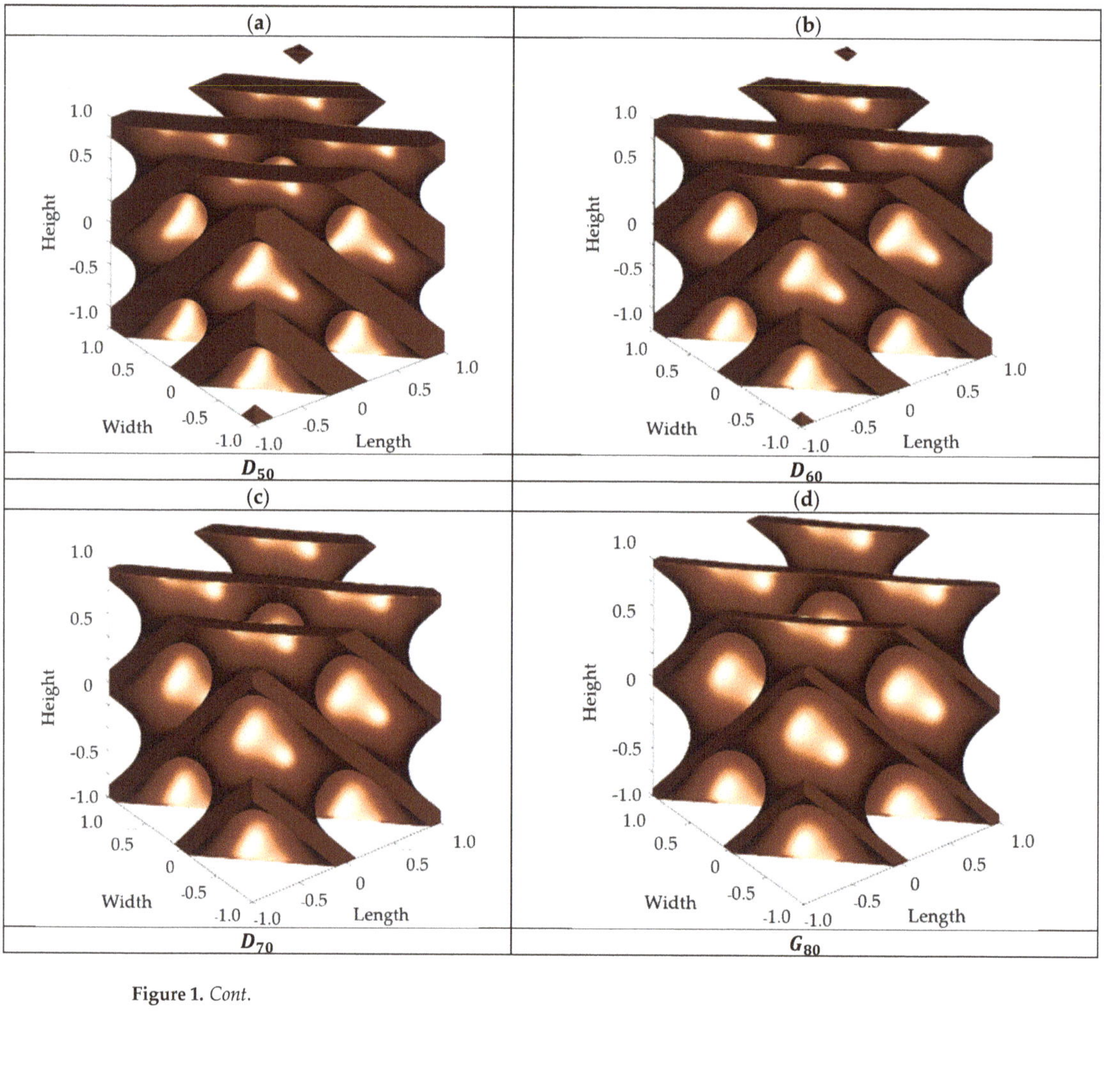

Figure 1. *Cont.*

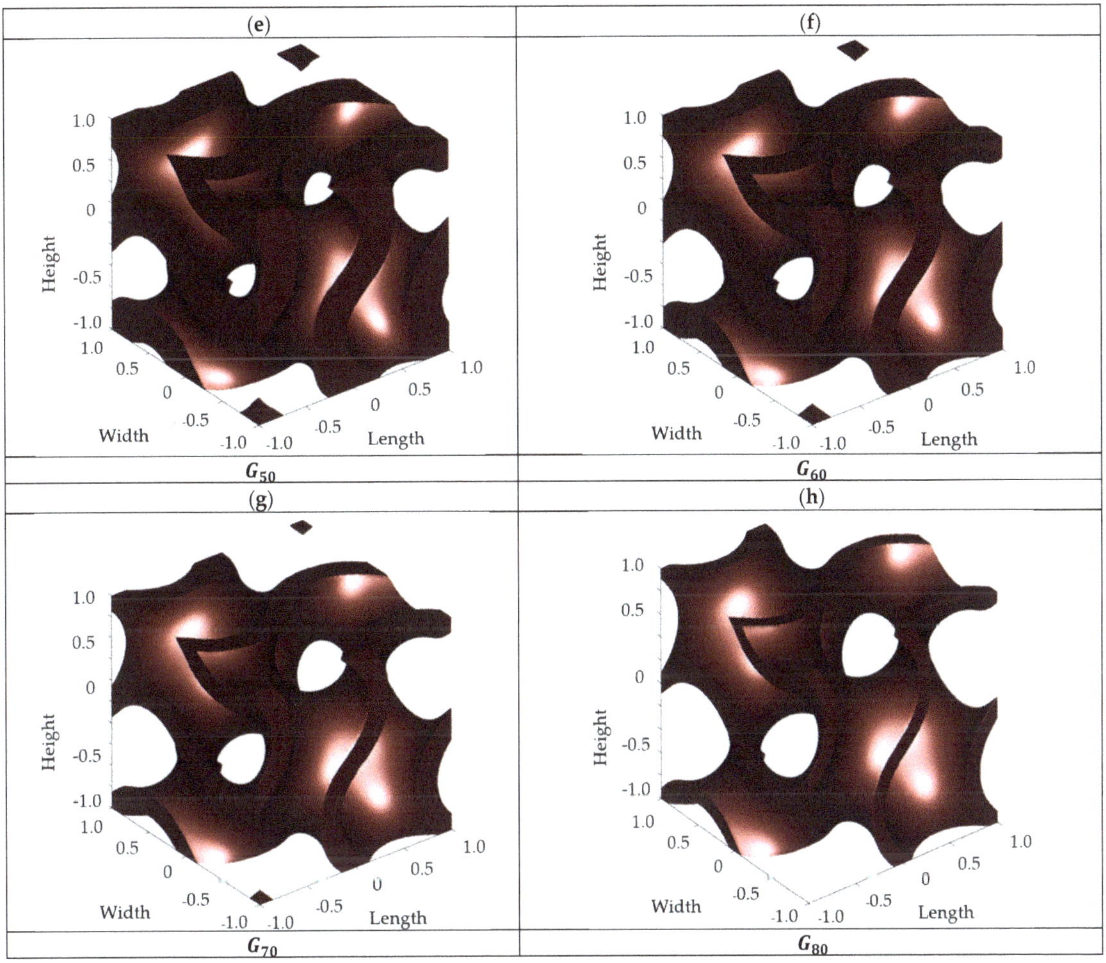

Figure 1. TPMS-based scaffold models with porosities of 50–80%: (**a–d**) Diamond; (**e–h**) Gyroid.

Table 1. The wall thickness of TPMS-based scaffolds.

TPMS Based Scaffolds	50%	60%	70%	80%
Diamond wall thickness (mm)	0.42	0.33	0.25	0.16
Gyroid wall thickness (mm)	0.51	0.41	0.31	0.21

2.2. Governing and Adopted CFD Equations

The following approach utilized the Navier–Stokes equation, taking into account an incompressible fluid with constant density.

$$\rho\frac{\partial u}{\partial t} - \mu\nabla^2 u + \rho(u.\nabla)u + \nabla p = F \tag{1}$$

$$\nabla.u = 0 \tag{2}$$

where ρ, u, μ, ∇, p, and F represent the density of the fluid $\left(\frac{kg}{m^3}\right)$, the fluid velocity (m/s), dynamic viscosity of fluid (kg/m.s), the del operator, pressure (Pa), and other forces (gravity or centrifugal force; in this case F = 0), respectively [6]. The blood density was assumed to remain constant at 1060 kg/m^3 for the CFD models.

The Carreau–Yasuda model was used to compute the fluid viscosity for the analysis, considering blood as a non-Newtonian fluid. The Carreau–Yasuda model was chosen for its ability to describe viscosity plateaus in bone scaffold applications, capturing fluid behavior under varying porosities. It proved essential in predicting pressure drop-flow rate relationships, demonstrating effectiveness in analyzing Carreau fluid flow through porous media [30]. The model's accuracy in predicting fluid behavior makes it a valuable tool for understanding and optimizing blood flow through intricate scaffold structures, aiding in the design of efficient bone tissue engineering systems.

$$\mu = \mu_\infty + (\mu_0 - \mu_\infty)\left[1 + \left(\lambda\dot{\gamma}\right)^2\right]^{(n-1)/2} \tag{3}$$

The equation utilized for Carreau fluids generalized from Darcy law is presented below:

$$\frac{\Delta p}{L} = \frac{U\eta_{eff}}{k} = \frac{U}{k}(\mu_\infty + (\mu_0 - \mu_\infty))\left[1 + (\lambda\dot{\gamma})^2\right]^{(n-1)/2} \tag{4}$$

where u and L denote the inlet fluid velocity (m/s) and model length (m), respectively. The values of μ_∞ and μ_0 represent the shear rate limit viscosities when the shear rate is either infinite or zero, respectively, relaxation time constant is λ, and the power law index is n [30].

In a laminar flow system, the WSS, denoted as $\tau\omega$, can be mathematically expressed as the normal velocity gradient on a wall.

$$\tau_\omega = \mu\frac{\partial u}{\partial m} \tag{5}$$

where m indicates the x, y, and z directions [16].

The Reynolds number for non-Newtonian fluid can be calculated by [31]:

$$R_e = \frac{\rho u L}{\mu_a} \tag{6}$$

where ρ, u, L, and μ_a denote density, velocity of fluid, length, and viscosity of fluid.

The Reynolds number for this non-Newtonian fluid under the given conditions is approximately 0.0052, which indicates a laminar flow regime. The characterization of blood flow as laminar in the CFD analysis of TPMS structures is based on physiological considerations, primarily the low flow velocities within blood vessels, resulting in Reynolds numbers conducive to laminar flow. The relatively high viscosity of blood and the small-scale nature of the vessels further support the assumption of laminar flow. This modeling choice aligns with typical flow conditions in the human circulatory system, enhancing the credibility of the simulation results.

2.3. Boundary Conditions

The study employed a laminar flow of fluid with an inlet velocity of 0.0001 m/s [6,17]. The direction of flow was chosen based on its potential to simplify the computational setup and boundary conditions, leading to a more accurate analysis and understanding of outcomes. Figure 2 shows the boundary conditions. To specify different fluid properties at the surface, such as velocity and pressure, a wall boundary condition was employed. In this study, a no-slip condition was imposed on the porous structure, which was assumed to be hydrophilic, consistent with typical practices in bone tissue engineering applications [25,32]. This assumption accounts for wetting behavior, leading to more accurate and practical CFD simulations. Furthermore, the outlet pressure was set to zero [15], to simplify the pressure differential computation, which was a critical parameter.

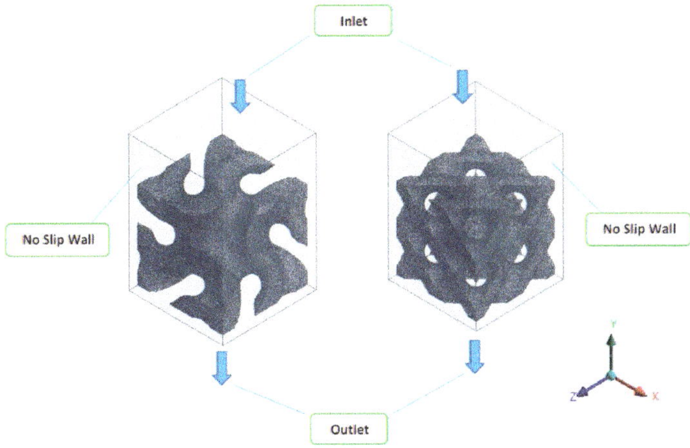

Figure 2. Boundary conditions for CFD analysis.

2.4. Numerical Schemes

In this study, the Finite Volume Method (FVM) has been employed as the numerical framework to simulate blood flow through the scaffold structures. To account for the non-Newtonian behavior of blood, the Carreau–Yasuda model has been integrated, offering a comprehensive description of viscosity variations with shear rate. The governing equations were solved using a coupled scheme, which simultaneously solves the momentum and continuity equations to enhance convergence stability and efficiency, particularly beneficial for flow interactions within the complex geometries of the scaffolds. For spatial discretization, the Green Gauss Node Based method was employed for gradient calculations due to its superior accuracy in handling unstructured meshes. The pressure interpolation was conducted using a second-order scheme to ensure a higher accuracy in capturing pressure variations within the flow domain. Momentum equations were discretized using a first-order upwind scheme, which, despite being less accurate than higher-order schemes, provides robust and stable solutions, particularly in the initial stages of simulation or in regions with high gradients and complex flow patterns. This combination of numerical schemes was chosen to balance computational efficiency and solution accuracy, ensuring reliable simulation results for the analysis of blood flow behavior through the TPMS scaffolds.

2.5. Grid Convergence

The study focused on the characterization of blood as a non-Newtonian fluid. A mesh sensitivity analysis was conducted to assess how variations in mesh element sizes influenced the results. Figure 3 shows a visual representation of the mesh in the fluid domain containing the scaffold. Mesh elements with sizes of 0.2, 0.5, and 1 mm were used to compare the permeability and WSS values of a specific model. The findings revealed that using a mesh size of 0.2 mm resulted in a minimal variation, less than 2%, in the WSS of the unit cell. As a result, a size of element of 0.2 mm was selected for the ongoing investigation. Furthermore, calculations were performed for WSS, permeability, pressure, and fluid velocity. The spatial discretization is accomplished using a tetrahedral structured mesh. To ensure the model's performance sensitivity, results were brought into convergence by establishing a residual criterion value of 1×10^{-6}.

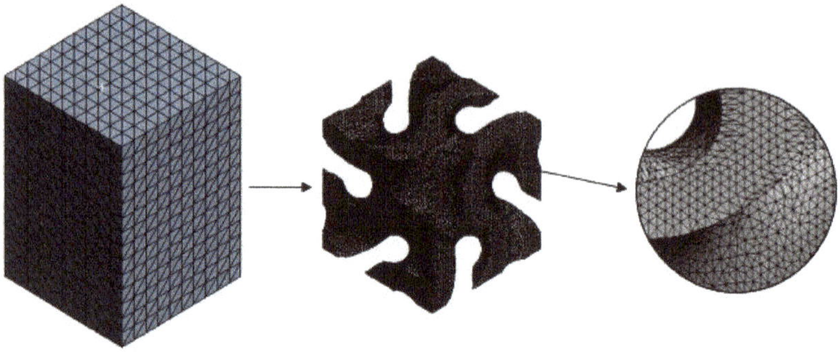

Figure 3. Visual representation of mesh in fluid domain containing the scaffold.

3. Numerical Model Validation

The present numerical methodology is validated by comparing the results with the literature. In this regard, the permeability results for the Gyroid structure are compared with the Ali and Sen [16] study. They analyzed the Gyroid structure using the power law model with an inlet velocity of 0.7 mm/s. The results from the present methodology match well with those from the Ali and Sen [16] study with a reasonable accuracy. Figure 4 shows the validation of our numerical methodology for the permeability of a Gyroid structure.

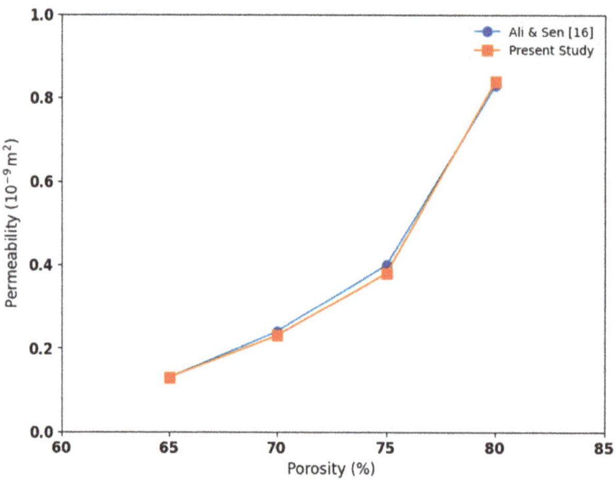

Figure 4. Validation of numerical methodology for the permeability of Gyroid structure [16].

4. Results and Discussions

4.1. Fluidic Characteristics of Scaffolds

An analysis of pressure and velocity distribution was conducted to examine the fluidic characteristics of Diamond (D) and Gyroid (G) structures with varying porosity levels ranging from 50% to 80%. The pressure distribution contours of the D surface (i.e., D_{50}, D_{60}, D_{70}, and D_{80}) and the G surface (i.e., G_{50}, G_{60}, G_{70}, and G_{80}) are presented in Figure 5. It is evident that there is a gradual decrease in pressure from the inlet to the outlet in all structures, regardless of the porosity level. The observed pressure variation in TPMS-based scaffolds, with the higher pressure near the inlet diminishing as the fluid progresses downwards, is a consequence of the intricate interplay between geometric complexities and fluid dynamics. The initial compression near the inlet, induced by the complex scaffold geometry, leads to regions of elevated pressure. Subsequent fluid flow through smoother paths and porous structures results in a gradual pressure dissipation. At the bottom of each picture, negative values of the pressure appear. This is because, near the outlet, the speed was increasing (Figure 5) so the pressure is decreased. Also, due to the effect of gravity, the speed increases as the fluid progresses so the pressure decreased.

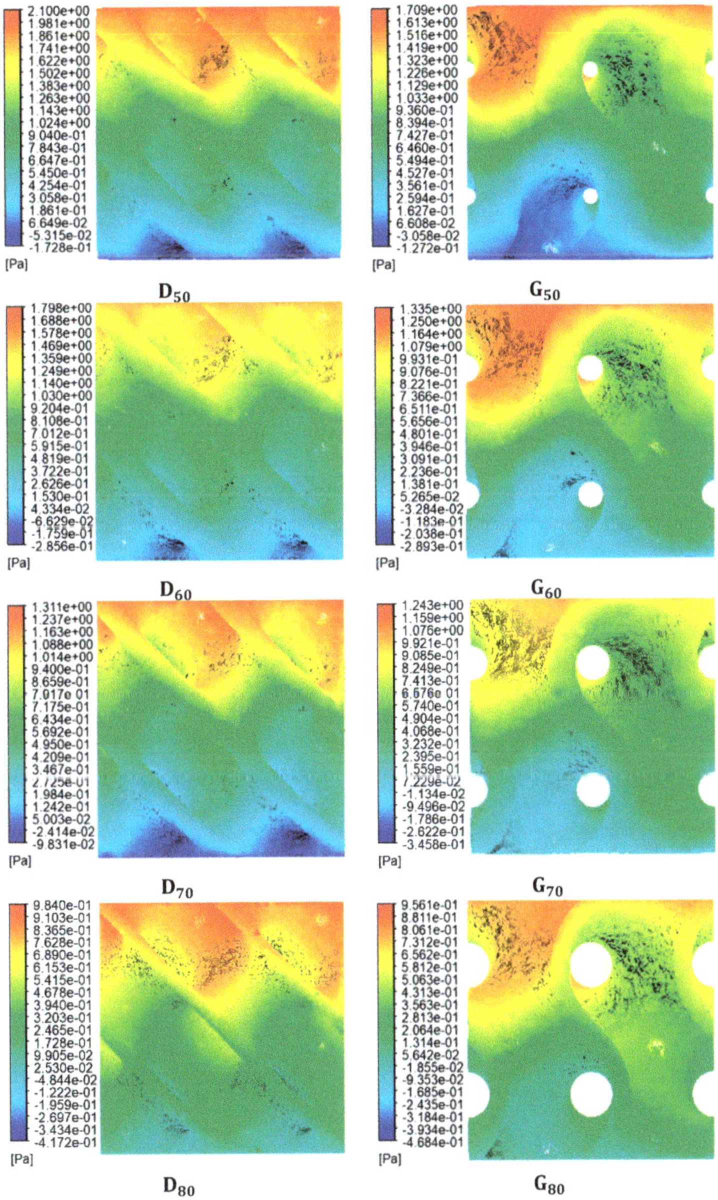

Figure 5. Pressure distribution contours of diamond structure (**left side**) and gyroid structures (**right side**).

The velocity distribution within scaffolds, as illustrated in Figure 6, reveals the profound impact of internal design on flow characteristics. It is crucial to observe that the central regions of scaffold structure surface have velocities that exceed the initial inlet velocity. The increase in speed can be attributed to the complex lattice geometry, which causes a localized rise in fluid velocity. At the same time, adjacent regions demonstrate the combined influence of geometric features and fluid dynamics [15]. The observed acceleration of the fluid profile, particularly in the central areas, has significant implications for the transportation of nutrients to deeper structures within the lattice. The improved fluid dynamics in these central regions suggest a potential optimization for the dispersion of nutrients, which agrees with results of the previous studies [33,34].

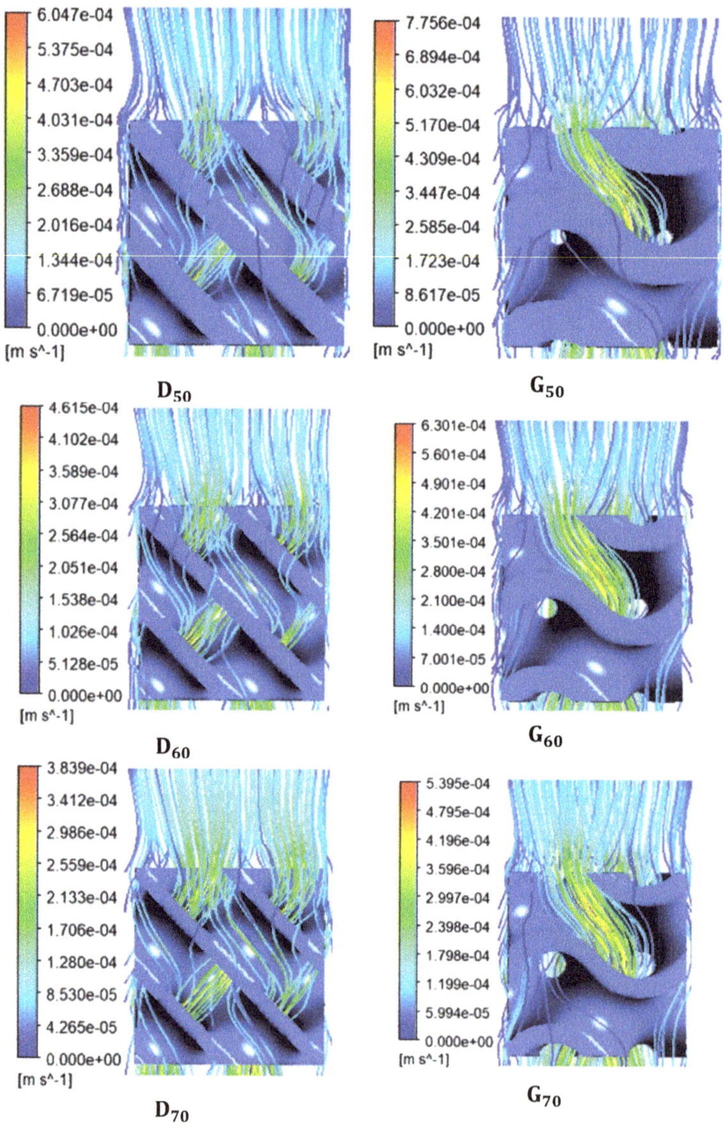

Figure 6. *Cont.*

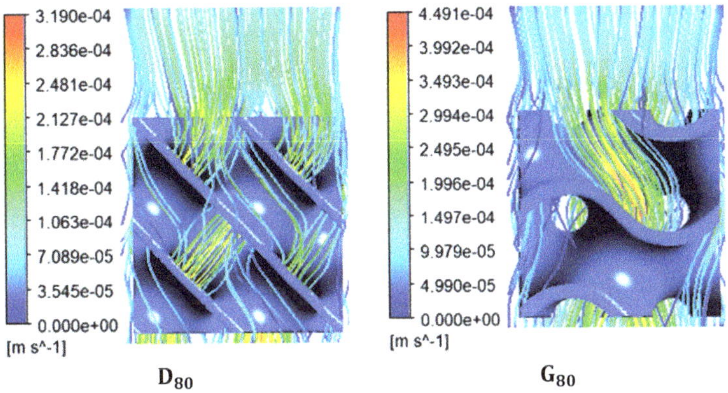

Figure 6. Velocity streamlines the distribution of diamond structures (**left side**) and gyroid structures (**right side**).

Figure 7 shows the pressure drop versus porosity, as by increasing the porosity, the pressure drop decreases in both the Diamond and Gyroid structures. The Diamond structure shows a higher pressure drop than the Gyroid structure due to its higher stiffness and strength, which result in a lower permeability. A study investigated the mechanical properties of TPMS-based Diamond and Gyroid Ti6Al4V scaffolds for bone implants. The findings revealed that, when comparing scaffolds with identical pore sizes, the Diamond scaffold exhibited a greater stiffness and strength than the Gyroid scaffold. As a result, the Diamond scaffold exhibited a higher pressure drop [35]. This phenomenon can be attributed to the inherent differences in the geometrical characteristics and flow dynamics of the two structures. The complex geometry of the Diamond structure induces more resistance to fluid flow, resulting in a higher pressure drop, whereas the Gyroid structure, with its distinct architecture, presents a comparatively lower resistance, leading to a reduced pressure drop. Figure 7 shows that the pressure drop ranges from 2.079 Pa to 0.984 Pa for the Diamond structure and 1.669 Pa to 0.943 Pa for the Gyroid structure for porosities ranging from 50% to 80%.

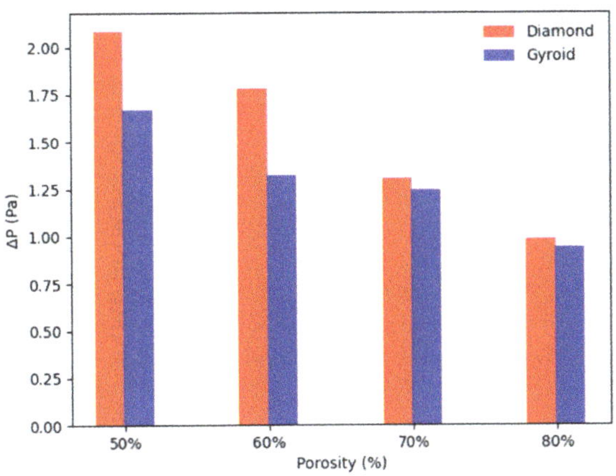

Figure 7. Pressure drop vs. porosity.

The TPMS structures are made from stainless steel, which has a high Young's modulus (200 GPa) and yield strength (500 MPa), making them rigid and resistant to deformation under fluid loads. Additionally, the small dimensions (a few millimeters) and thick walls of the structures ensure structural integrity, minimizing deformation. Similar studies, such as those by Song et al. [36], have also assumed non-deformable vessels for TPMS structures due to their rigidity and structural design, supporting our assumption.

4.2. Permeability

Equation (4) was used to calculate the permeability of the Diamond and Gyroid structures, considering the pressure difference obtained through the CFD analysis. To assess the transport efficiency of different porous structures, both permeability (k) and pressure difference (Δp) were taken into account.

For the successful occurrence of mass transfer within any porous structure, permeability is of the utmost importance. In Figure 8, the expected permeability of both the Diamond and Gyroid surfaces is depicted for the porosity range of 50% to 80%. It is noted that a higher porosity results in an increased permeability because the pores in the scaffold allow for better fluid flow. Research has indicated that TPMS scaffolds with higher porosity levels have higher permeability. The Gyroid scaffold type has been determined to be more permeable than the Diamond scaffold type [37]. The continuous, three-dimensional architecture of the Gyroid offers a larger surface area for the flow of fluids and facilitates a more interconnected pathway for fluid passage. These characteristics of the Gyroid, including its enhanced interconnectivity and greater void space, contribute to its higher permeability, allowing fluids to move through the Gyroid scaffold with reduced resistance compared to the Diamond scaffold. Ali et al. [23] studied Gyroid- and lattice-based architectures with porosities ranging from 65% to 90% to observe their permeability. Our results show a good agreement with the results from this study. Gómez S. et al. [38] observed the properties of 3D scaffolds for bone tissue engineering. They showed that the properties of scaffolds can be controlled in design stage to match exactly with trabecular bone. Van Bael S. [39] performed a study to observe the effect of pore size and permeability on Ti6Al4V scaffolds of three different pore shapes at 500 μm and 1000 μm. Nauman E. et al. [40] conducted a study to observe the permeability of trabecular bone for tissue engineering applications. Beaudoin et al. [41] observed the permeability of cancellous bone.

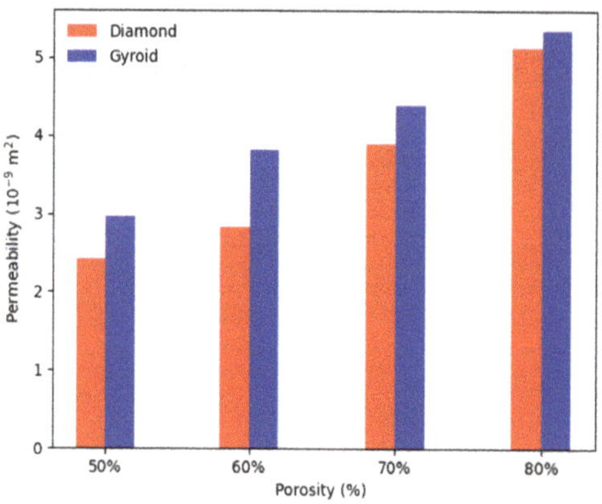

Figure 8. Permeability vs. porosity.

In this study, the permeability ranged from 2.424×10^{-9} m^2 to 5.122×10^{-9} m^2 for the Diamond structure and 2.966×10^{-9} m^2 to 5.344×10^{-9} m^2 for the Gyroid structure. Table 2 shows the good agreement of the obtained results with previously reported results.

Table 2. Permeability of porous scaffolds and bone.

	Ali D. et al. [23]	Gómez S. et al. [38]	Van Bael S. et al. [39]	Nauman E. A. et al. [40]	Beaudoin A. J. et al. [41]	Current Work
Permeability (10^{-9} m^2)	1.0–36.0 #	5.0–45.0 #	5.06–30.50 #	0.0268–20 *	0.467–14.80 *	2.424–5.122

Porous scaffold #; Bone *.

4.3. WSS

Equation (5) was utilized to determine the WSS of diamond and gyroid structures. The analysis of the WSS on the Gyroid and Diamond structures revealed a consistent trend. WSS consistently decreased with an increase in porosity for both types of structures, which attributed to the reduced surface area available for fluid interaction within the porous framework.

When comparing the WSS values between the two TPMS structures, the Diamond TPMS exhibited consistently lower WSS values compared to the Gyroid TPMS across various porosities shown in Figure 9. The results show that the Diamond TPMS is more effective in reducing WSS compared to the Gyroid TPMS, which is beneficial for improving the overall performance of the TPMS structure. This difference in WSS values between the two TPMS structures may have significant implications for their performance in different applications. Figure 9 shows that the WSS ranges from 9.903×10^{-2} Pa to 9.840×10^{-1} Pa for the Diamond structure and 1.150×10^{-1} Pa to 7.717×10^{-2} Pa for the Gyroid structure.

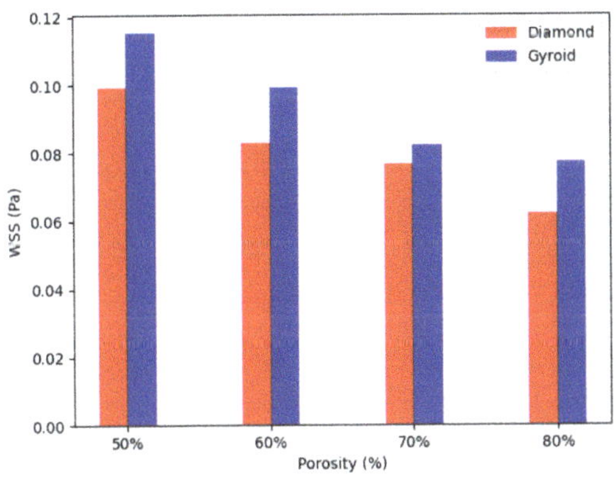

Figure 9. WSS vs. porosity.

5. Conclusions

In this study, the fluidic properties of Diamond and Gyroid TPMS-based scaffolds with varying porosities (50–80%) were examined through CFD analysis, focusing on their potential applications in orthopedic implants. The findings highlight significant differences in pressure distribution, permeability, and WSS between the two structures. Gyroid scaffolds consistently demonstrated superior fluidic performance compared to Diamond structures, with lower pressure drops and higher permeability across all porosity levels. These characteristics suggest that Gyroid scaffolds are more conducive to nutrient flow and waste removal, and their gradual decrease in WSS with increased porosity indicates a favorable environment for cell proliferation and tissue regeneration. Therefore, Gyroid structures are recommended as the preferred choice for orthopedic implants applications due to their balanced fluidic properties and potential to support biological processes effectively.

Although the results were derived from a model with limited dimensions (2 mm × 2 mm × 2 mm), they provide foundational insights that could be applicable to larger scaffold samples. Future research should focus on scaling these models and validating them with experimental data to ensure their applicability to clinically relevant scaffold sizes. The necessity of experimental validation is evident; while the simulations offer a comprehensive understanding of fluidic behavior, physical experiments, such as flow perfusion bioreactor studies, are required to confirm the predicted pressure drops, permeability, and WSS values under physiological conditions. Additionally, experimental studies could explore the biological responses, such as cell adhesion and growth, within these scaffolds under dynamic flow conditions.

Author Contributions: Conceptualization, M.N.S., M.U.S., S.R., M.I. and M.A.O.; formal analysis, M.N.S., M.U.S., S.R., M.I. and M.A.O.; funding acquisition, S.R. and M.A.O.; investigation, M.N.S., M.U.S., S.R. and M.I.; methodology, M.N.S., M.U.S. and S.R.; project administration, S.R.; supervision, S.R. and M.I.; writing—original draft, M.N.S. and M.U.S.; writing—review and editing, S.R. and M.I. All authors have read and agreed to the published version of the manuscript.

Funding: This research received no external funding.

Data Availability Statement: The original contributions presented in the study are included in the article, further inquiries can be directed to the corresponding author.

Conflicts of Interest: The authors declare no conflicts of interest.

References

1. Yadav, P.; Beniwal, G.; Saxena, K.K. A Review on Pore and Porosity in Tissue Engineering. *Mater. Today Proc.* **2021**, *44*, 2623–2628. [CrossRef]
2. Rasheed, S.; Lughmani, W.A.; Obeidi, M.A.; Brabazon, D.; Ahad, I.U. Additive Manufacturing of Bone Scaffolds Using Polyjet and Stereolithography Techniques. *Appl. Sci.* **2021**, *11*, 7336. [CrossRef]
3. Ali, D.; Sen, S. Computational Fluid Dynamics Study of the Effects of Surface Roughness on Permeability and Fluid Flow-Induced Wall Shear Stress in Scaffolds. *Ann. Biomed. Eng.* **2018**, *46*, 2023–2035. [CrossRef]
4. Rasheed, S.; Lughmani, W.A.; Khan, M.M.; Brabazon, D.; Obeidi, M.A.; Ahad, I.U. The Porosity Design and Deformation Behavior Analysis of Additively Manufactured Bone Scaffolds through Finite Element Modelling and Mechanical Property Investigations. *J. Funct. Biomater.* **2023**, *14*, 496. [CrossRef]
5. Castro, A.P.G.; Pires, T.; Santos, J.E.; Gouveia, B.P.; Fernandes, P.R. Permeability versus Design in TPMS Scaffolds. *Materials* **2019**, *12*, 1313. [CrossRef]
6. Kumar, J.; Nirala, N.S.; Singh, N.K.; Gupta, N.; Dwivedi, Y.D.; Verma, R.; Rai, S.K.; Gupta, M. Design, Development and Fluidic Behavior Analysis of Triply Periodic Minimal Surface (TPMS) Based Scaffolds for Bone-Applications. *Int. J. Interact. Des. Manuf.* **2023**, 1–11. [CrossRef]
7. Wang, Z.; Huang, C.; Wang, J.; Wang, P.; Bi, S.; Abbas, C.A. Design and Simulation of Flow Field for Bone Tissue Engineering Scaffold Based on Triply Periodic Minimal Surface. *Chin. J. Mech. Eng.* **2019**, *32*, 19. [CrossRef]
8. Poltue, T.; Karuna, C.; Khrueaduangkham, S.; Seehanam, S.; Promoppatum, P. Design Exploration of 3D-Printed Triply Periodic Minimal Surface Scaffolds for Bone Implants. *Int. J. Mech. Sci.* **2021**, *211*, 106762. [CrossRef]
9. Abbasi, N.; Hamlet, S.; Love, R.M.; Nguyen, N.T. Porous Scaffolds for Bone Regeneration. *J. Sci. Adv. Mater. Devices* **2020**, *5*, 1–9. [CrossRef]
10. Restrepo, S.; Ocampo, S.; Ramirez, J.A.; Paucar, C.; Garcia, C. Mechanical Properties of Ceramic Structures Based on Triply Periodic Minimal Surface (TPMS) Processed by 3D Printing. *J. Phys. Conf. Ser.* **2017**, *935*, 012036. [CrossRef]
11. Barba, D.; Alabort, E.; Reed, R.C. Synthetic Bone: Design by Additive Manufacturing. *Acta Biomater.* **2019**, *97*, 637–656. [CrossRef] [PubMed]
12. Rahbari, A.; Montazerian, H.; Davoodi, E.; Homayoonfar, S. Predicting Permeability of Regular Tissue Engineering Scaffolds: Scaling Analysis of Pore Architecture, Scaffold Length, and Fluid Flow Rate Effects. *Comput. Methods Biomech. Biomed. Engin* **2017**, *20*, 231–241. [CrossRef] [PubMed]
13. Chabanon, M.; Duval, H.; Grenier, J.; Beauchesne, C.; Goyeau, B.; David, B. Histological Method to Study the Effect of Shear Stress on Cell Proliferation and Tissue Morphology in a Bioreactor. *Tissue Eng. Regen. Med.* **2019**, *16*, 225–235. [CrossRef] [PubMed]
14. Pires, T.; Santos, J.; Ruben, R.B.; Gouveia, B.P.; Castro, A.P.G.; Fernandes, P.R. Numerical-Experimental Analysis of the Permeability-Porosity Relationship in Triply Periodic Minimal Surfaces Scaffolds. *J. Biomech.* **2021**, *117*, 110263. [CrossRef]
15. Ma, S.; Tang, Q.; Han, X.; Feng, Q.; Song, J.; Setchi, R.; Liu, Y.; Liu, Y.; Goulas, A.; Engstrøm, D.S.; et al. Manufacturability, Mechanical Properties, Mass-Transport Properties and Biocompatibility of Triply Periodic Minimal Surface (TPMS) Porous Scaffolds Fabricated by Selective Laser Melting. *Mater. Des.* **2020**, *195*, 109034. [CrossRef]
16. Ali, D.; Sen, S. Permeability and Fluid Flow-Induced Wall Shear Stress of Bone Tissue Scaffolds: Computational Fluid Dynamic Analysis Using Newtonian and Non-Newtonian Blood Flow Models. *Comput. Biol. Med.* **2018**, *99*, 201–208. [CrossRef] [PubMed]
17. Ali, D.; Ozalp, M.; Blanquer, S.B.G.; Onel, S.; Permeability, S.O. Permeability and Fluid Flow-Induced Wall Shear Stress in Bone Scaffolds with TPMS and Lattice Architectures: A CFD Analysis. *Eur. J. Mech. B/Fluids* **2020**, *79*, 376–385. [CrossRef]
18. Hollister, S.J. Porous Scaffold Design for Tissue Engineering. *Nat. Mater.* **2005**, *4*, 518–524. [CrossRef] [PubMed]

19. Aqila, N.; Hussein, K.; Aziah, N.; Mohtar, J.; Haqim Neza, L.; Noordin, M.A.; Putra, A.; Saad, M. Pore Size Effect on Mechanical Response and Fluid Permeability for Bone Scaffold Regeneration. *Mal. J. Med. Health Sci.* **2021**, *17*, 34–39.
20. Ntousi, O.; Roumpi, M.; Siogkas, P.; Deligianni, D.; Fotiadis, D.I. Computational Fluid Dynamic Analysis of Customised 3D-Printed Bone Scaffolds with Different Architectures. In Proceedings of the 2023 45th Annual International Conference of the IEEE Engineering in Medicine & Biology Society (EMBC), Sydney, Australia, 24–27 July 2023; Volume 2023. [CrossRef]
21. Guyot, Y.; Papantoniou, I.; Luyten, F.P.; Geris, L. Coupling Curvature-Dependent and Shear Stress-Stimulated Neotissue Growth in Dynamic Bioreactor Cultures: A 3D Computational Model of a Complete Scaffold. *Biomech. Model. Mechanobiol.* **2016**, *15*, 169–180. [CrossRef]
22. Maskery, I.; Sturm, L.; Aremu, A.O.; Panesar, A.; Williams, C.B.; Tuck, C.J.; Wildman, R.D.; Ashcroft, I.A.; Hague, R.J.M. Insights into the Mechanical Properties of Several Triply Periodic Minimal Surface Lattice Structures Made by Polymer Additive Manufacturing. *Polymer* **2018**, *152*, 62–71. [CrossRef]
23. Ali, D.; Sen, S. Finite Element Analysis of Mechanical Behavior, Permeability and Fluid Induced Wall Shear Stress of High Porosity Scaffolds with Gyroid and Lattice-Based Architectures. *J. Mech. Behav. Biomed. Mater.* **2017**, *75*, 262–270. [CrossRef] [PubMed]
24. Lesman, A.; Blinder, Y.; Levenberg, S. Modeling of Flow-Induced Shear Stress Applied on 3D Cellular Scaffolds: Implications for Vascular Tissue Engineering. *Biotechnol. Bioeng.* **2010**, *105*, 645–654. [CrossRef]
25. Marin, A.C.; Lacroix, D. The Inter-Sample Structural Variability of Regular Tissue-Engineered Scaffolds Significantly Affects the Micromechanical Local Cell Environment. *Interface Focus* **2015**, *5*, 20140097. [CrossRef] [PubMed]
26. Zhao, F.; Vaughan, T.J.; McNamara, L.M. Quantification of Fluid Shear Stress in Bone Tissue Engineering Scaffolds with Spherical and Cubical Pore Architectures. *Biomech. Model. Mechanobiol.* **2016**, *15*, 561–577. [CrossRef]
27. Lee, H.G.; Park, J.; Yoon, S.; Lee, C.; Kim, J. Mathematical Model and Numerical Simulation for Tissue Growth on Bioscaffolds. *Appl. Sci.* **2019**, *9*, 4058. [CrossRef]
28. Lu, Y.; Zhao, W.; Cui, Z.; Zhu, H.; Wu, C. The Anisotropic Elastic Behavior of the Widely-Used Triply-Periodic Minimal Surface Based Scaffolds. *J. Mech. Behav. Biomed. Mater.* **2019**, *99*, 56–65. [CrossRef] [PubMed]
29. Sewify, G.H.; Javid, K.; Adeel, M.; Abbasi, A.; Khan, S.U.; Omri, M.; Kolsi, L. Blood Flow in Multi-Sinusoidal Curved Passages with Biomimetic Rheology: An Application of Blood Pumping. *Mathematics* **2022**, *10*, 1579. [CrossRef]
30. Shahsavari, S.; McKinley, G.H. Mobility of Power-Law and Carreau Fluids through Fibrous Media. *Phys. Rev. E Stat. Nonlin Soft Matter Phys.* **2015**, *92*, 063012. [CrossRef]
31. Cho, Y. Effects of the Non-Newtonian Viscosity of Blood on Hemodynamics of Diseased Arterial Flows. *Biorheology* **1991**, *28*, 241–262. [CrossRef]
32. Truscello, S.; Kerckhofs, G.; Van Bael, S.; Pyka, G.; Schrooten, J.; Van Oosterwyck, H. Prediction of Permeability of Regular Scaffolds for Skeletal Tissue Engineering: A Combined Computational and Experimental Study. *Acta Biomater.* **2012**, *8*, 1648–1658. [CrossRef] [PubMed]
33. Ma, S.; Tang, Q.; Feng, Q.; Song, J.; Han, X.; Guo, F. Mechanical Behaviours and Mass Transport Properties of Bone-Mimicking Scaffolds Consisted of Gyroid Structures Manufactured Using Selective Laser Melting. *J. Mech. Behav. Biomed. Mater.* **2019**, *93*, 158–169. [CrossRef] [PubMed]
34. Jain, T.; Jain, J.K.; Saxena, K.K. Design and Comprehensive Study of Biodegradable Zinc-Based Implants for Bio-Medical Applications. *Adv. Mater. Process. Technol.* **2022**, *8*, 519–536. [CrossRef]
35. Naghavi, S.A.; Tamaddon, M.; Marghoub, A.; Wang, K.; Babamiri, B.B.; Hazeli, K.; Xu, W.; Lu, X.; Sun, C.; Wang, L.; et al. Mechanical Characterisation and Numerical Modelling of TPMS-Based Gyroid and Diamond Ti6Al4V Scaffolds for Bone Implants: An Integrated Approach for Translational Consideration. *Bioengineering* **2022**, *9*, 504. [CrossRef] [PubMed]
36. Song, J.; Wang, M.; Li, D.; Zhang, J. Deformation and Energy Absorption Performance of Functionally Graded TPMS Structures Fabricated by Selective Laser Melting. *Appl. Sci.* **2024**, *14*, 2064. [CrossRef]
37. Santos, J.; Pires, T.; Gouveia, B.P.; Castro, A.P.G.; Fernandes, P.R. On the Permeability of TPMS Scaffolds. *J. Mech. Behav. Biomed. Mater.* **2020**, *110*, 103932. [CrossRef] [PubMed]
38. Gómez, S.; Vlad, M.D.; López, J.; Fernández, E. Design and Properties of 3D Scaffolds for Bone Tissue Engineering. *Acta Biomater.* **2016**, *42*, 341–350. [CrossRef] [PubMed]
39. Van Bael, S.; Chai, Y.C.; Truscello, S.; Moesen, M.; Kerckhofs, G.; Van Oosterwyck, H.; Kruth, J.P.; Schrooten, J. The Effect of Pore Geometry on the in Vitro Biological Behavior of Human Periosteum-Derived Cells Seeded on Selective Laser-Melted Ti6Al4V Bone Scaffolds. *Acta Biomater.* **2012**, *8*, 2824–2834. [CrossRef]
40. Nauman, E.A.; Fong, K.E.; Keaveny, T.M. Dependence of Intertrabecular Permeability on Flow Direction and Anatomic Site. *Ann. Biomed. Eng.* **1999**, *27*, 517–524. [CrossRef]
41. Beaudoin, A.J.; Mihalko, W.M.; Krause, W.R. Finite Element Modelling of Polymethylmethacrylate Flow through Cancellous Bone. *J. Biomech.* **1991**, *24*, 127–136. [CrossRef]

designs

Article

Optimizing Selective Laser Melting of Inconel 625 Superalloy through Statistical Analysis of Surface and Volumetric Defects

Ali Shahrjerdi [1,*], Mojtaba Karamimoghadam [2], Reza Shahrjerdi [3], Giuseppe Casalino [2] and Mahdi Bodaghi [4,*]

1. Mechanical Engineering Department, Malayer University, Malayer 7R7X+8M5, Iran
2. Department of Mechanics, Mathematics and Management, Polytechnic University of Bari, Via Orabona 4, 70125 Bari, Italy; m.karamimoghadam@phd.poliba.it (M.K.); giuseppe.casalino@poliba.it (G.C.)
3. Department of Industrial, Mechanical and Aerospace Engineering, Buin Zahra Technical and Engineering University, Buin Zahra Q346+GVC, Iran; r.shahrjerdi@bzte.ac.ir
4. Department of Engineering, School of Science and Technology, Nottingham Trent University, Nottingham NG11 8NS, UK

* Correspondence: alishahrjerdi2000@yahoo.com (A.S.); mahdi.bodaghi@ntu.ac.uk (M.B.)

Abstract: This article delves into optimizing and modeling the input parameters for the selective laser melting (SLM) process on Inconel 625. The primary aim is to investigate the microstructure within the interlayer regions post-process optimization. For this study, 100 layers with a thickness of 40 μm each were produced. Utilizing the design of experiments (DOE) methodology and employing the Response Surface Method (RSM), the SLM process was optimized. Input parameters such as laser power (LP) and hatch distance (HD) were considered, while changes in microhardness and roughness, Ra, were taken as the responses. Sample microstructure and surface alterations were assessed via scanning electron microscopy (SEM) analysis to ascertain how many defects and properties of Inconel 625 can be controlled using DOE. Porosity and lack of fusion, which were due to rapid post-powder melting solidification, prompted detailed analysis of the flaws both on the surfaces of and in terms of the internal aspects of the samples. An understanding of the formation of these imperfections can help refine the process for enhanced integrity and performance of Inconel 625 printed material. Even slight directional changes in the columnar dendrite structures are discernible within the layers. The microstructural characteristics observed in these samples are directly related to the parameters of the SLM process. In this study, the bulk samples achieved a microhardness of 452 HV, with the minimum surface roughness recorded at 9.9 μm. The objective of this research was to use the Response Surface Method (RSM) to optimize the parameters to result in the minimum surface roughness and maximum microhardness of the samples.

Keywords: additive manufacturing; selective laser melting; Inconel 625 superalloy; roughness; microhardness

1. Introduction

Today, the use of additive manufacturing (AM) in the production of complex components has become very common. Many intricate metal parts, in terms of their geometry, are accurately and rapidly fabricated using the SLM method [1–4]. Although employing this technique requires process optimization, engineers have striven to optimize input parameters such as laser power, laser speed, hatch distance, and other input variables through various methods [5,6] to optimize concise indexes like the volumetric energy density [7]. Metal powders like Inconel are utilized to manufacture very delicate components in the medical, petrochemical, and automotive industries. After production, these powders might undergo changes, leading to defects such as a lack of fusion, cracks, porosities, and other manufacturing flaws, which can be mitigated by optimizing the input parameters [8–10].

Comprehensive studies on processing Inconel 625 alloy via SLM have explored its microstructure, mechanical properties, and process optimization. Characterization using SEM, EBSD, and XRD revealed columnar crystals and high hardness due to rapid cooling.

Heat treatment induced distinct microstructures and altered the grain boundaries [11]. Nickel-based superalloy powders were assessed, linking their properties to the SLM parameters and the mechanical behavior in Inconel 625 parts [12]. Aerospace applications were examined, comparing SLM and laser metal deposition (LMD) methods, highlighting differences in the microstructure and material properties. The microstructural evolution post-SLM and heat treatment clarified the uncertainties in the mechanical properties [10]. Investigations into the impact of the SLM parameters revealed specific track characteristics and microstructural changes, while studies on lattice structures aimed to understand their mechanical behavior and validate numerical models [13].

Predictive methods for the temperature profiles and melt pool shapes in SLM were developed [14,15]. These studies emphasized microstructural variations, mechanical responses, and optimization strategies crucial for enhancing Inconel 625's properties and applications in various industries [16–19]. Kundakcıoğlu et al. [20] introduced a thermal model predicting transient temperatures and molten pool shapes, which they validated experimentally. Sun et al. [21] investigated high-temperature oxidation behavior, linking lower laser energy densities to increased oxidation due to hindered protective oxide layer formation. Balbaa et al. [22] explored the SLM parameters' impact on part properties, emphasizing density, surface quality, and residual stresses. Shrestha et al. [23] analyzed porosity and single-track geometry variations based on SLM parameters, while Mazur et al. [24] sought defect reduction strategies in SLM fabrication. Yan et al. [25] studied the build orientation's influence on the surface structure and tribological properties. Additionally, studies by [26,27] compared the microstructural and mechanical properties under different manufacturing conditions, emphasizing the impacts of mechanical anisotropy, the printing directions, and welding speed on the material characteristics in SLM-manufactured Inconel 625. Huang et al. [28] investigated the scanning speed's influence on oxidation resistance and mechanical properties and attributed superior performance to unique grain boundaries. The effects of strain rate on Inconel 625's mechanical behavior, explored by Du et al. [29], included rate sensitivity and temperature-induced softening. Zhang et al. [30] examined the impacts of microstructural aspects and heat treatment on Inconel 625 alloy blades and noted changes in the crystal grains and enhanced corrosion resistance. Pleass et al. [31] delved into SLM's effects on grain structures and mechanical properties, highlighting anisotropic microstructures' role in enhancing failure resistance. Other studies investigated heat treatment's impact on the deformation mechanisms, identified core–shell-structured oxides, and analyzed the tensile strength under various AM parameters, with the aim of optimizing Inconel 625's mechanical properties. Additionally, research has focused on thermo-fluid conditions in laser surface melting, alloying of Inconel 625 at different scanning speeds, and computational models' accuracy in predicting the melt pool geometry. The impact of annealing on the microstructure and corrosion resistance was elucidated for various factors affecting the mechanical properties and behavior of selectively laser-melted Inconel 625 [32–37]. Hu et al. [38] investigated the microstructure of Inconel 625 using DED, highlighting significant anisotropy in its mechanical properties and fatigue crack growth, attributed to the orientation of its epitaxial columnar grains. During heat treatment of Inconel 625 [39], changes were noted in its grain structure and phase dissolution, which impacted its strength and ductility differently based on the treatment methods. Li et al. [40] developed a precise 3D finite element heat transfer model for SLM with Inconel 625, integrating various heat source models, achieving a higher accuracy with a hybrid model, and enhancing the simulation's precision. Allam et al. [41] explored phased array ultrasonic testing (PAUT) for non-destructive evaluation of SLM parts, successfully detecting their internal porosity using multiple methods. Wormald et al. [42] provided solutions for predicting cross-sectional dimensions in AM, achieving high accuracy in predicting track dimensions. Teng et al. [43] investigated SLM's impact on Inconel 718, identifying heat treatment-induced phase changes, defect formations, and anisotropic mechanical properties, offering potential for tailored tensile properties. Schmeiser et al. [44] documented microstructural evolution in laser powder bed fusion

(LPBF) of Inconel 625, studying how process parameters influence texture, defects, recrystallization, and segregation. They noted distinct texture generation influenced by laser power and scanning speed, along with solid-state texture changes without remelting, linking them to in situ recrystallization and subsequent segregation growth. Zhou et al. [45] developed a novel hybrid scanning strategy that offers theoretical guidance for restoration processes. However, limited research restricts further LPBF technology development. This study comprehensively investigates the hybrid scanning strategy's impact on the microstructure, mechanical properties, and interfacial characteristics of Inconel 625, providing significant insights for designing additively manufactured Inconel 625 with specific microstructures and properties. Accurate prediction of temperature distribution and melting pool geometry in additive manufacturing requires careful selection of heat source models. Binxun et al. [46] compares surface, volumetric, and double-ellipsoid models, implemented in Abaqus/Standard. Results show the double-ellipsoid model best predicts melting pool geometry, while the volumetric model is better for peak temperatures. Surface model accuracy is poor, emphasizing the need for model calibration.

The crucial aspect lies in uncovering the correlation between process parameters and defects, as it has the potential to reduce both time and costs associated with the manufacturing process. The present study aims to explore the optimization of the SLM parameters for Inconel 625, focusing on interlayer microstructure post-process. This research tries to optimize parameters using RSM to achieve minimum surface roughness and maximum microhardness in the samples. Using DOE and RSM, we optimized SLM parameters like laser power, and hatch distance, which are correlated to microhardness and roughness. Examining cubic Inconel 625 samples, it revealed porosities and fusion issues due to rapid solidification, prompting a detailed analysis for process enhancement and material performance. The highest microhardness reported for the bulk samples is 452 HV with a roughness of 10.2 μm. This was achieved using a laser power of 175 W and a hatch distance of 0.1 mm.

2. Response Surface Methodology

Utilizing optimization techniques is crucial for selecting top-quality samples while saving time and costs. RSM is particularly adept at streamlining SLM by organizing both input and output parameters [47,48]. Its objective is to create an accurate mathematical model for the SLM process, minimizing errors within Equation (1) where η is considered a response and k steps are controlled factors. Ultimately, RSM aims to optimize the printing process efficiently and effectively.

$$\eta = f(x_1, x_2, \ldots, x_k) + \varepsilon \quad (1)$$

The variable ε represents the unpredictable SLM error stemming from uncontrollable factors. Accurately establishing the η response is pivotal; a closer approximation of η to the actual value enhances the reliability of the function on various factors. Equation (2) presents another formula aimed at determining the true response factor for η.

$$\eta = \beta_0 + \sum_{i=1}^{k} \beta_i x_i + \sum_{i=1}^{k} \beta_{ii} x_i^2 + \sum_{i.j=1} \sum_{i<j} \beta_{ij} x_i x_j \quad (2)$$

β_0 represents a constant, β_i signifies a linear factor, and β_{ii} denotes an interaction factor. The study focused on LP and HD as controlled factors, utilizing statistical analysis through Design-Expert V13. In total, nine samples were manufactured, and the assessment focused on three optimal samples, considering three different output response levels. In the DOE, the experimental design was chosen according to the varying levels of these input parameters. The study examined a set of experiments designed to correlate specific input settings with desired output responses, aiming to identify optimal sample configurations.

3. Experimental Work

In this experiment, cubes with dimensions of 10 × 10 × 4 mm^2 were produced by SLM method due to setup Figure 1a, totaling nine samples based on DOE Table 1. A thickness of 40 μm was considered for each sample (Figure 1b,c), consisting of 100 layers fabricated using the SLM method with Inconel 625 powder without reinforcement, as specified in Table 2. SEM images were taken by Tescan Mira3 SEM (Tescan Co., Brno, Czechia) in Backscattered electrons mood. For surface roughness measurement, Alicona InfiniteFocus SL (Bruker Alicona, Raaba, Austria) was used to measure Ra on the X direction of the surface. To measure the microhardness, a test that involved loading 200 gf via Micromet No. 5114 (BUEHLER micro-durometer, Neuss, Germany) was applied for five repetitions in each area.

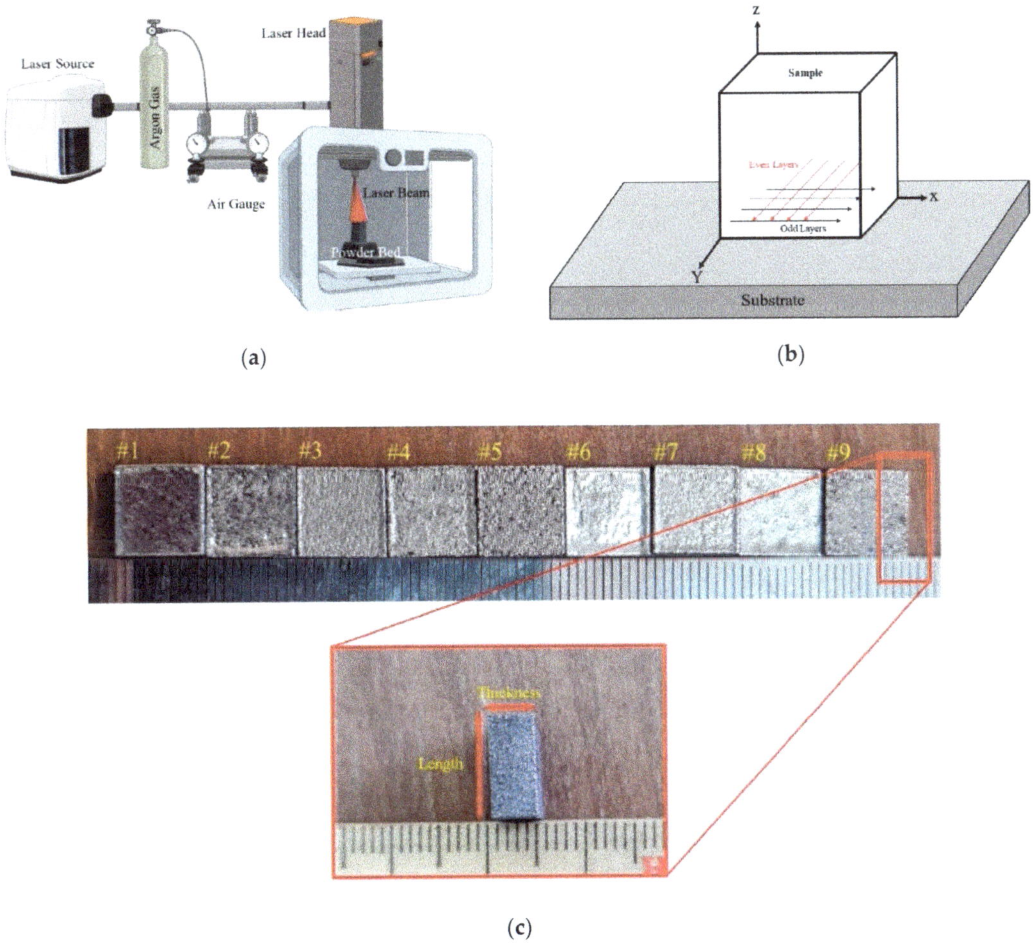

Figure 1. (**a**) Schematic of selective laser melting for processing Inconel 625 powder (**b**) sample images of #1 to #9 (**c**) dimensions of the samples for length (10 mm) and thickness (4 mm).

Table 1. Design of experiments' input and output parameters.

No.	Input Parameters		Responses	
	Laser Power (W)	Hatch Distance (mm)	Microhardness (hv)	Roughness (μm)
1	200	0.1	448	9.9
2	150	0.1	449	10.3
3	200	0.3	425	11.9
4	175	0.2	422	11.5
5	200	0.2	436	10.7
6	150	0.3	406	13.2
7	175	0.3	418	12.5
8	150	0.2	408	11.8
9	175	0.1	452	10.2

Table 2. Chemical composition of Inconel 625 powder.

Elements	Ni	Fe	Mo	Nb	Co	Ta	Ti	Cr	Al
%	60	5	10	0.9	1	0.8	0.7	21	0.6

4. Discussion

4.1. Microstructure Analysis of AM Samples

Figure 2 displays the microstructure of Inconel 625 in samples #1, #3, and #7 post-additively manufactured through SLM. In Figure 2a, the deposited layers exhibit distinct zones, which were influenced both by a high cooling rate and process direction, demonstrating the influence of manufacturing process factors in the SLM method. This variance in the columnar dendrite deposition structure impacts microhardness, which is evident in Table 1, where the higher the LS the lower the microhardness. The centralized pattern in Figure 2b indicates a characteristic solidification behavior potentially linked to laser scan strategies and energy input. Furthermore, the merging of melting pools in Figure 2c implies complex interactions between laser settings, impacting the fusion and morphology of adjacent layers. This detailed analysis underscores the sensitivity of Inconel 625's microstructure to variations in SLM parameters, emphasizing the need for precise control and optimization to tailor material characteristics for specific applications. Figure 3a shows the macro scale of the deposited layer after the SLM process, while Figure 3b illustrates the outcome of rapid solidification, depicting a centralized pattern post-process with all columnar dendrite directions converging at a single center. Additionally, Figure 2c highlights instances where reduced HD and increased LP cause certain melting pools to merge in select zones, hinting at alterations within the heat-affected zone (HAZ). It is crucial to grasp the significance of the overlap percentage in relation to HAZ. Even minor directional adjustments in columnar dendrite structures are noticeable within the layers. The microstructural features observed in these samples bear direct relevance to the SLM process parameters. The zone within layers suggests varying cooling rates during deposition, which impacts dendritic growth and ultimately influences material properties like microhardness.

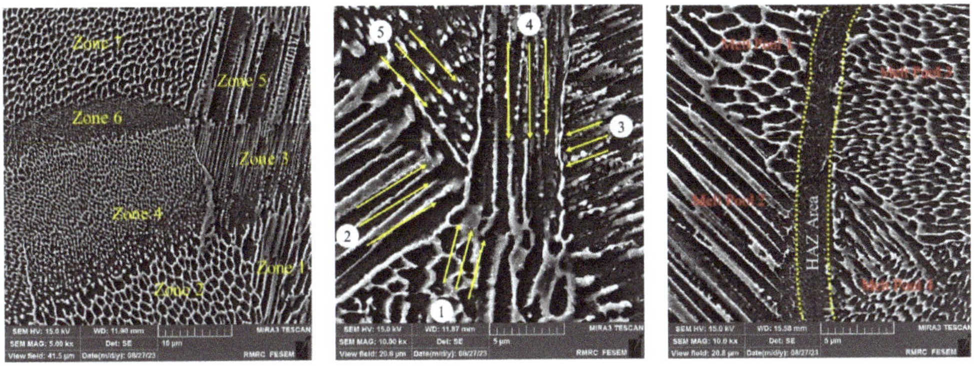

(a) (b) (c)

Figure 2. Microstructure analysis of samples #1, #3, and #7 with different input parameters including LP = 150–200 W, and HD = 0.1–0.3 mm (**a**) sample #1, (**b**) sample #3, and (**c**) sample #7.

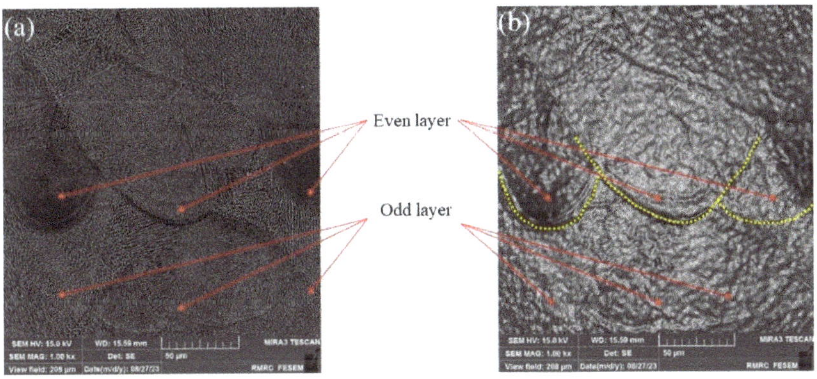

Figure 3. Macro structures of SLM sample #1 considering (LP = 200 W, and HD = 0.1 mm) the process parameters.

4.2. Revealing Defects

Lack of Fusion and Pores after SLM Process

Insufficient fusion at the interfaces of five adjacent layers in Inconel 625 SLM samples, which is particularly noticeable when four layers are formed consecutively, is often observed. This lack of fusion results in the formation of voids and pores (Figure 4). LP and powder properties significantly impact interlayer fusion. Variety in these parameters affects the molten pool temperature distribution and the material cooling rates, which influences the fusion quality between layers. The rapid solidification during SLM can hinder proper bonding, leading to pores or voids at the interfaces. Inconel 625 microstructure analysis corroborates this hypothesis, revealing irregularities and insufficient bonding at interfaces and further pore formation. Understanding and optimizing the interaction between process parameters and microstructural characteristics are crucial for mitigating this lack of fusion and enhancing material integrity in SLM-produced components. Inconel 625 sample exhibits open pores, particularly clustered along the curves of melting pools, with some in close proximity to each other. The presence of open pores in Inconel 625 samples manufactured through SLM can be linked to several causes. During the SLM process, the rapid heating and subsequent rapid cooling rates can lead to incomplete fusion or inadequate time for gas to escape, which leads to gases being entrapped in pockets and thus leads to the formation of open pores. The formation of open pores near the curves of melting pools could be related to the keyhole effect in laser melting. This effect involves the rapid vaporization of material under the intense heat of the laser, which can create voids or open pores along the edges of the melted pools. Inconel 625 powder properties, such as particle size distribution and morphology, can impact pore formation. Agglomerated or irregularly shaped powder particles might not fuse properly during the melting process, leading to open pores.

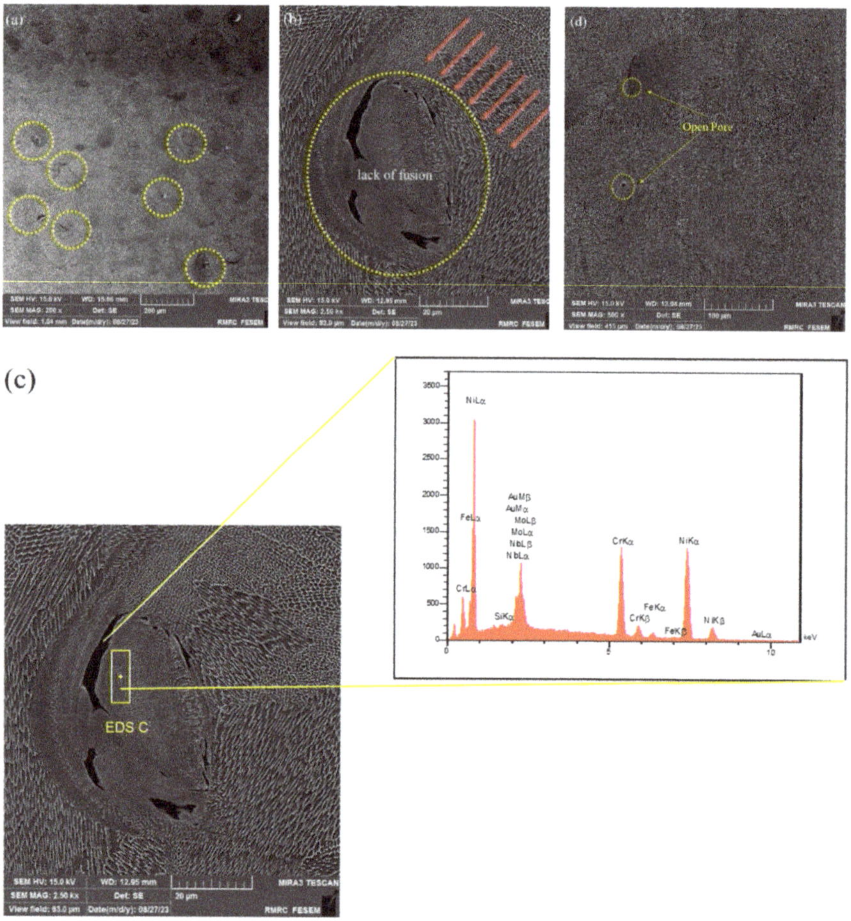

Figure 4. The defects on samples #5, #8, and #11 (**a**) porosities on sample #5, (**b**,**c**) lack of fusion of sample #8. (**d**) Open pore on sample #11.

4.3. Surface Roughness, Ra (SR)

Table 3 illustrates the ANOVA outcomes for SR. It reveals the effectiveness of all main linear and quadratic parameters. Within the quadratic terms, the laser power's quadratic term (P2) stands out significantly, alongside three impactful interactions depicted in the table. Table 3 shows the analysis of variance ANOVA model of SR. Equations (3) and (4) display the regression relationship for SR, considering the influential terms based on coded and actual values.

$$\text{Roughness, Ra} = +423.11 + 7.33\ \text{Laser Power} - 16.33\ \text{B} + 4.50\ \text{Laser Power} \times \text{Hatch Distance} - 1.67\ \text{Laser Power}^2 + 11.33\ \text{Hatch Distance}^2 \quad (3)$$

$$\text{Roughness, Ra} = +431.11111 + 0.866667\ \text{Laser Power} - 931.66667\ \text{Hatch Distance} + 1.80000\ \text{Laser Power} \times \text{Hatch Distance} - 0.002667\ \text{Laser Power}^2 + 1133.33333\ \text{Hatch Distance}^2 \quad (4)$$

Table 3. Analysis of variance ANOVA model of surface roughness.

Source	Sum of Squares	df	Mean Square	F-Value	*p*-Value	
Model	2266.78	5	453.36	9.90	0.0440	significant
A-Laser Power	322.67	1	322.67	7.04	0.0767	
B-Hatch Distance	1600.67	1	1600.67	34.94	0.0097	
Laser Power × Hatch Distance	81.00	1	81.00	1.77	0.2757	
Laser Power^2	5.56	1	5.56	0.1213	0.7507	
Hatch Distance^2	256.89	1	256.89	5.61	0.0987	
Residual	137.44	3	45.81			
Cor Total	2404.22	8				

Figure 5a shows the response surface graph, while Figure 5b displays a contour plot for SR concerning LP, and HD. Two-dimensional-representation contour plots of the surface graphs delineate the parameter effectiveness zones for each response more clearly. These graphs, based on regression, are interpolated by the available data. The interaction of hatch distance and LP on SR is evident in Figure 5a, showing SR dependency on parameter changes. For instance, decreasing LP and increasing HD escalate surface roughness. This occurs because reduced LP may hinder adequate energy delivery for complete material melting, leading to incomplete fusion between layers and, consequently, surface irregularities. Likewise, widened hatch distances limit overlap between laser paths, resulting in visible gaps and rougher surfaces (Figure 6). This combination fosters incomplete melting and inadequate fusion, culminating in heightened surface roughness in the SLM process. Exposing the samples to air and cooling them after fabrication by SLM led to the formation of surface pores due to heat transfer to the substrate and air. These pores resulted in a rougher surface.

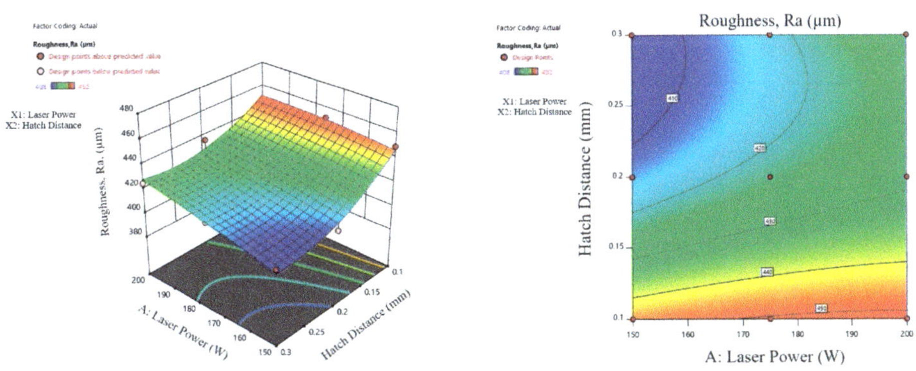

(a) (b)

Figure 5. (**a**) Surface roughness diagrams of response surface graph of LP and HD. (**b**) contour plot of HD and LP.

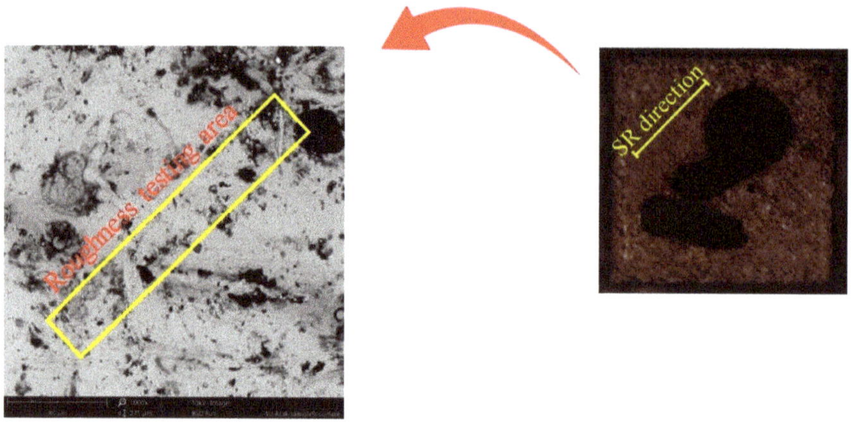

Figure 6. Surface appearance of the direction of surface roughness test on sample #9 by considering Ra factor for the measurement.

Figure 7 shows various plots from a statistical analysis; in more detail, Figure 7a presents the normal plot of residuals. This plot illustrates that the differences between observed and predicted values in the regression analysis are tightly clustered around the line, indicating a close alignment. Additionally, Figure 7b graphically demonstrates the proximity of actual and predicted amounts to the line, which accounts for the accurate prediction by the statistical analysis of the responses. Figure 7c shows the perturbation plot of SR, highlighting the depiction of parameter interactions affecting SR and the simultaneous impact of these parameters on SR. By comparing Figures 5 and 7c, it is evident that the SR significantly decreased as the HD increased. This reduction occurred because the gap was more effectively filled in the layers fabricated at higher HD, leading to more uniform metal solidification.

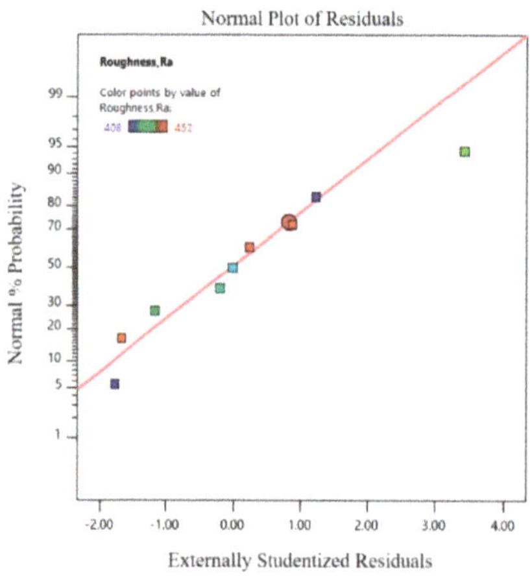

(a)

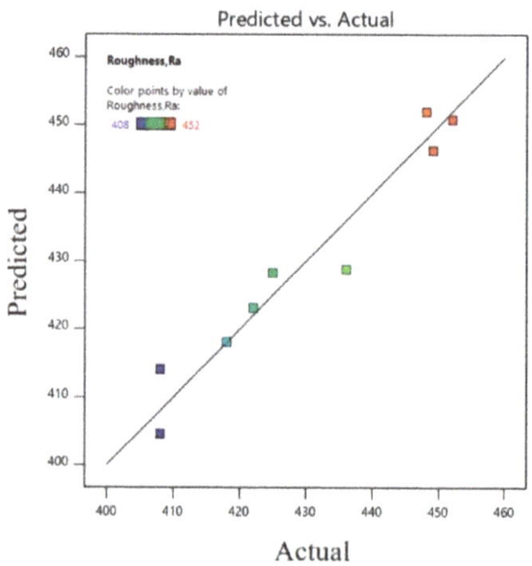

(b)

Figure 7. *Cont.*

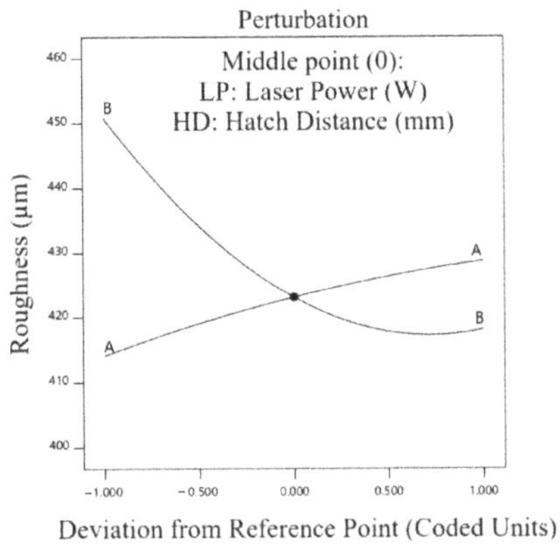

(c)

Figure 7. (**a**) Normal plot of residuals; (**b**) predicted and actual graph; (**c**) perturbation plot.

4.4. Microhardness Profiles

Table 4 illustrates the ANOVA outcomes for microhardness. It reveals the effectiveness of all main linear and quadratic parameters. Within the quadratic terms, the LP and HD quadratic terms (A^2, B^2, and C^2) stand out significantly, alongside three impactful interactions depicted in the table. Analyzing ANOVA Table 4, Equations (5) and (6) display the regression relationship for the microhardness, considering the influential terms based on coded and actual values, respectively. Since the model was highly significant, with a reported p-value of 0.0012, it was presented in the form shown in Table 4 to account for the quadratic term.

$$\text{Microhardness} = +11.40 - 0.4667\ \text{A} + 1.20\ \text{Hatch Distance} - 0.2250\ \text{Laser Power} \times \text{Hatch Distance} - 0.1000\ \text{Laser Power}^2 + 0.0000\ \text{Hatch Distance}^2 \quad (5)$$

$$\text{Microhardness} = +4.21667 + 0.055333\ \text{Laser Power} + 27.75000\ \text{Hatch Distance} - 0.090000\ \text{Laser Power} \times \text{Hatch Distance} - 0.000160\ \text{Laser Power}^2 - 1.81542 \times -13\ \text{Hatch Distance}^2 \quad (6)$$

Table 4. Analysis of variance ANOVA model of microhardness.

Source	Sum of Squares	df	Mean Square	F-Value	p-Value	
Model	10.17	5	2.03	120.03	0.0012	significant
A-Laser Power	1.31	1	1.31	77.11	0.0031	
B-Hatch Distance	8.64	1	8.64	509.90	0.0002	
AB	0.2025	1	0.2025	11.95	0.0407	
A^2	0.0200	1	0.0200	1.18	0.3568	
B^2	0.0000	1	0.0000	0.0000	1.0000	
Residual	0.0508	3	0.0169			
Cor Total	10.22	8				

Figure 8a shows the response surface graph, while Figure 8b displays a contour plot for microhardness concerning LP and HD parameters. These graphs, based on regression Equations (5) and (6), are interpolated by the data. Due to the coefficients of each parameter in the regression equations and static analysis, the plots related to the interaction effect of LP with HD, have assumed an oscillatory pattern. This phenomenon arises from the rapid freezing and solidification of the powder following its melting by the laser within each distinct interval. The microhardness of the samples undergoes variations contingent upon these conditions, which are highly influenced by input parameters.

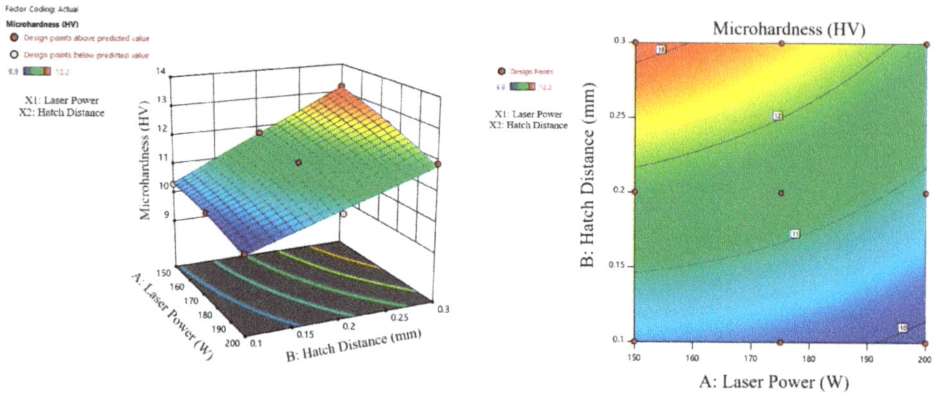

(**a**) (**b**)

Figure 8. (**a**) Microhardness diagrams of response surface graph of LP and HD. (**b**) Contour plot of LP and HD.

Figure 9 displays the normal plot of residuals. This plot illustrates that the differences between observed and predicted values in the regression analysis are tightly clustered around the line, indicating a close alignment. Additionally, Figure 9b graphically demonstrates the proximity of actual and predicted amounts to the line, signifying the accurate prediction by the statistical analysis of the responses for microhardness analysis. Figure 9c shows the perturbation plot of microhardness, highlighting the depiction of parameter interactions affecting microhardness and the simultaneous impact of these parameters on microhardness. As shown in Figure 9c, increasing the laser power (LP) led to an increase in microhardness due to the proper melting of the powder in the melt pool, resulting in the complete formation of the metal.

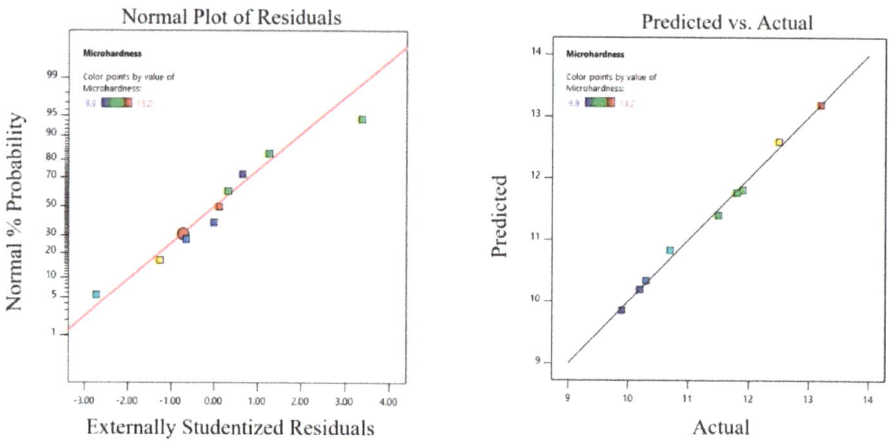

(**a**) (**b**)

Figure 9. *Cont.*

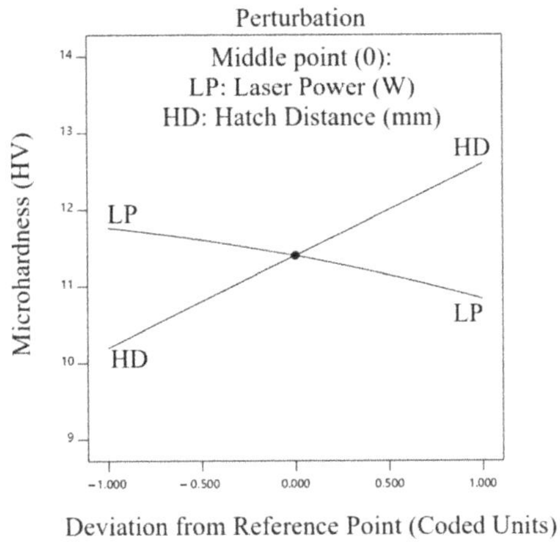

(c)

Figure 9. (**a**) Normal plot of residuals; (**b**) predicted and actual graph; (**c**) perturbation plot.

5. Optimization

Based on statistical analysis of the regression model of relations between input parameters and output responses, optimization was performed using the RSM method. The statistical analysis of experimental data led to the regression equations that depict the connections between input parameters and output responses. Employing the function within the Design Expert V13 software facilitated process optimization to achieve specific conditions. Table 5 lists the constraints and criteria for both variables and responses to optimize the SLM process. Within Table 5, three solutions are presented, aiming to optimize the SLM process based on the criteria of minimizing SR and maximizing microhardness. All parameters and responses were given equal consideration, as the statistical analysis identified the significance of each input and output parameter in both the SR and microhardness sections.

Table 5. Constraints of SLM input and output parameters.

Parameter/Response		Goal	Lower Limit	Upper Limit	Importance
Parameters	LP (W)	in range	150	200	3
	HD (mm)	in range	0.1	0.3	3
Response	SR (μm)	Minimize	9.7	13.2	3
	Microhardness (hv)	Maximize	400	500	3

After conducting the software analysis, three optimal samples were selected for potential modification to achieve two distinct responses (as shown in Table 6). This process involves a comprehensive review of the input parameters and their corresponding responses, which are illustrated in the overlay plots in Figure 10. The yellow areas in these plots represent the optimal response levels. This implies that selecting process parameters from the gray areas would result in insufficient microhardness and surface roughness. It is important to note that the software, aiming to minimize surface roughness, identified 10.1–10.4 μm as an acceptable range, while for microhardness, the maximum range selected was 432–438 HV. The software determined the range of input and output parameters based on DOE Table 1.

Table 6. Predicted and real parameters for optimum samples.

	Parameters			Response	
No.	Laser Power (W)	Hatch Distance (mm)		SR (μm)	Microhardness (hv)
1	200	0.300	Predicted	10.1	438
2	200	0.300	Predicted	10.2	435
3	190	0.200	Predicted	10.4	432

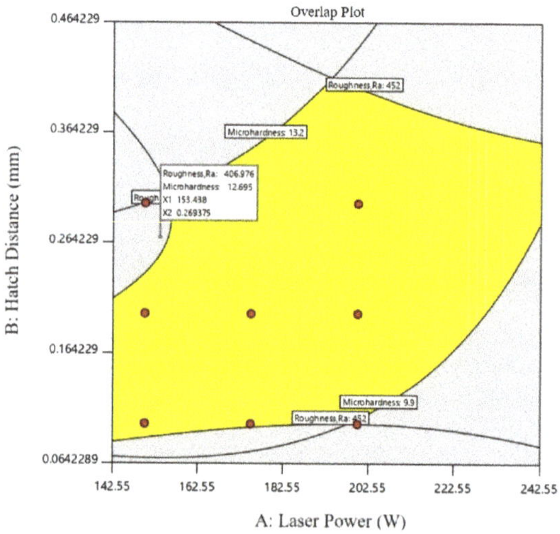

Figure 10. Overlay contour plot of hatch distance and laser power. The yellow area shows the optimum area and the gray area shows the less effective process parameters of the SLM process for Inconel 625.

6. Conclusions

In conclusion, this study extensively explores the optimization of SLM parameters for Inconel 625, focusing on microstructure analysis and some defects after. Employing DOE and RSM, the investigation considered the correlation between LP and HD with microhardness and SR responses. SEM analysis revealed surface defects, guiding improvements for enhanced material integrity and performance. The subsequent bullet list shows the most significant findings from this investigation:

- Insufficient fusion at the interfaces of five adjacent layers in Inconel 625 SLM samples, particularly when formed consecutively, often results in voids or pores within the material, impacting fusion quality.
- Open pores in Inconel 625 SLM samples are linked to factors like rapid heating and cooling rates, keyhole effect, and powder properties, emphasizing the need to optimize process parameters for enhanced material integrity.
- Response surface graphs and contour plots for SR parameters are derived from interpolated data using regression equations, providing clearer delineation of parameter effectiveness zones.
- The interaction between HD and LP significantly influences SR, with decreasing LP and increasing HD leading to escalated roughness. Reduced LP impedes adequate energy delivery for complete material melting, resulting in incomplete fusion and visible gaps between laser paths, fostering incomplete melting and inadequate fusion.

- The maximum microhardness achieved was 452 hv using an LP of 175 W and a hatch distance of 0.1 mm, while the minimum SR was 9.9 μm with an LP of 200 W and the same 0.1 mm HD.
- As part of the optimization process, three samples were introduced, with LP levels ranging between 190 and 200 W and HD values between 0.2 and 0.3 mm.

Author Contributions: Conceptualization, M.K. and A.S.; methodology, M.K., M.B., A.S. and G.C.; software, M.K.; validation, M.K., A.S., M.B., R.S. and G.C.; investigation, M.K.; resources, A.S.; writing—original draft preparation, M.K., R.S. and M.B.; writing—review and editing, A.S., M.K., R.S., M.B and G.C.; supervision, A.S.H, and G.C.; project administration, A.S. All authors have read and agreed to the published version of the manuscript.

Funding: This research received no external funding.

Data Availability Statement: All data are included in the article.

Conflicts of Interest: The authors declare that they have no conflicts of interest.

Abbreviations

SLM	Selective laser melting.
DOE	Design of experiments.
RSM	Response Surface Method.
LP	Laser power.
HD	Hatch distance.
SR	Surface roughness.
AM	Additive manufacturing.
LMD	Laser metal deposition.
LPBF	Laser powder bed fusion.
SEM	Scanning Electron Microscope.
HAZ	Heat-affected zone.
ANOVA	Analysis of variance.

References

1. Ahmad, M.N.; Yahya, A. Effects of 3D printing parameters on mechanical properties of ABS samples. *Designs* **2023**, *7*, 136. [CrossRef]
2. Blakey-Milner, B.; Gradl, P.; Snedden, G.; Brooks, M.; Pitot, J.; Lopez, E.; Leary, M.; Berto, F.; Du Plessis, A. Metal additive manufacturing in aerospace: A review. *Mater. Des.* **2021**, *209*, 110008. [CrossRef]
3. Vafadar, A.; Guzzomi, F.; Rassau, A.; Hayward, K. Advances in metal additive manufacturing: A review of common processes, industrial applications, and current challenges. *Appl. Sci.* **2021**, *11*, 1213. [CrossRef]
4. Li, Y.; Gu, D. Parametric analysis of thermal behavior during selective laser melting additive manufacturing of aluminum alloy powder. *Mater. Des.* **2014**, *63*, 856–867. [CrossRef]
5. Casalino, G.; Campanelli, S.L.; Contuzzi, N.; Ludovico, A.D. Experimental investigation and statistical optimisation of the selective laser melting process of a maraging steel. *Opt. Laser Technol.* **2015**, *65*, 151–158. [CrossRef]
6. Fang, Z.C.; Wu, Z.L.; Huang, C.G.; Wu, C.W. Review on residual stress in selective laser melting additive manufacturing of alloy parts. *Opt. Laser Technol.* **2020**, *129*, 106283. [CrossRef]
7. Caiazzo, F.; Alfieri, V.; Casalino, G. On the Relevance of volumetric energy density in the investigation of inconel 718 laser powder bed fusion. *Materials* **2020**, *13*, 538. [CrossRef]
8. Akram, J.; Pal, D.; Stucker, B. Establishing flow stress elongation relationships as a function of microstructural features of Ti6Al4V alloy processed using SLM. *Designs* **2019**, *3*, 21. [CrossRef]
9. Chen, F.; Wang, Q.; Zhang, C.; Huang, Z.; Jia, M.; Shen, Q. Microstructures and mechanical behaviors of additive manufactured Inconel 625 alloys via selective laser melting and laser engineered net shaping. *J. Alloys Compd.* **2022**, *917*, 165572. [CrossRef]
10. Marchese, G.; Garmendia Colera, X.; Calignano, F.; Lorusso, M.; Biamino, S.; Minetola, P.; Manfredi, D. Characterization and comparison of Inconel 625 processed by selective laser melting and laser metal deposition. *Adv. Eng. Mater.* **2017**, *19*, 1600635. [CrossRef]
11. Li, S.; Wei, Q.; Shi, Y.; Zhu, Z.; Zhang, D. Microstructure characteristics of Inconel 625 superalloy manufactured by selective laser melting. *J. Mater. Sci. Technol.* **2015**, *31*, 946–952. [CrossRef]
12. Pleass, C.; Jothi, S. Influence of powder characteristics and additive manufacturing process parameters on the microstructure and mechanical behaviour of Inconel 625 fabricated by Selective Laser Melting. *Addit. Manuf.* **2018**, *24*, 419–431. [CrossRef]

13. Li, C.; White, R.; Fang, X.Y.; Weaver, M.; Guo, Y.B. Microstructure evolution characteristics of Inconel 625 alloy from selective laser melting to heat treatment. *Mater. Sci. Eng. A* **2017**, *705*, 20–31. [CrossRef]
14. Li, C.; Guo, Y.B.; Zhao, J.B. Interfacial phenomena and characteristics between the deposited material and substrate in selective laser melting Inconel 625. *J. Mater. Process. Technol.* **2017**, *243*, 269–281. [CrossRef]
15. Leary, M.; Mazur, M.; Williams, H.; Yang, E.; Alghamdi, A.; Lozanovski, B.; Zhang, X.; Shidid, D.; Farahbod-Sternahl, L.; Witt, G.; et al. Inconel 625 lattice structures manufactured by selective laser melting (SLM): Mechanical properties, deformation and failure modes. *Mater. Des.* **2018**, *157*, 179–199. [CrossRef]
16. Criales, L.E.; Arısoy, Y.M.; Özel, T. Sensitivity analysis of material and process parameters in finite element modeling of selective laser melting of Inconel 625. *Int. J. Adv. Manuf. Technol.* **2016**, *86*, 2653–2666. [CrossRef]
17. Wang, P.; Zhang, B.; Tan, C.C.; Raghavan, S.; Lim, Y.F.; Sun, C.N.; Wei, J.; Chi, D. Microstructural characteristics and mechanical properties of carbon nanotube reinforced Inconel 625 parts fabricated by selective laser melting. *Mater. Des.* **2016**, *112*, 290–299. [CrossRef]
18. Hu, X.A.; Zhao, G.L.; Liu, F.C.; Liu, W.X. Microstructure mechanical behavior of Inconel 625 alloy processed by selective laser melting at high temperature up to 1000 °C. *Rare Met.* **2020**, *39*, 1181–1189. [CrossRef]
19. Fang, X.Y.; Li, H.Q.; Wang, M.; Li, C.; Guo, Y.B. Characterization of texture and grain boundary character distributions of selective laser melted Inconel 625 alloy. *Mater. Charact.* **2018**, *143*, 182–190. [CrossRef]
20. Kundakcıoğlu, E.; Lazoglu, I.; Poyraz, Ö.; Yasa, E.; Cizicioğlu, N. Thermal and molten pool model in selective laser melting process of Inconel 625. *Int. J. Adv. Manuf. Technol.* **2018**, *95*, 3977–3984. [CrossRef]
21. Sun, Y.; Chen, L.; Li, L.; Ren, X. High-temperature oxidation behavior and mechanism of Inconel 625 super-alloy fabricated by selective laser melting. *Opt. Laser Technol.* **2020**, *132*, 106509. [CrossRef]
22. Balbaa, M.A.; Elbestawi, M.A.; McIsaac, J. An experimental investigation of surface integrity in selective laser melting of Inconel 625. *Int. J. Adv. Manuf. Technol.* **2019**, *104*, 3511–3529. [CrossRef]
23. Shrestha, S.; Chou, K. An investigation into melting modes in selective laser melting of Inconel 625 powder: Single track geometry and porosity. *Int. J. Adv. Manuf. Technol.* **2021**, *114*, 3255–3267. [CrossRef]
24. Mazur, M.; Benoit, M.; Easton, M.; Brandt, M. Selective laser melting of Inconel 625 alloy with reduced defect formation. *J. Laser Appl.* **2020**, *32*, 022058. [CrossRef]
25. Yan, X.; Gao, S.; Chang, C.; Huang, J.; Khanlari, K.; Dong, D.; Ma, W.; Fenineche, N.; Liao, H.; Liu, M. Effect of building directions on the surface roughness, microstructure, and tribological properties of selective laser melted Inconel 625. *J. Mater. Process. Technol.* **2021**, *288*, 116878. [CrossRef]
26. Liu, M.; Wang, Q.; Cai, Y.; Lu, D.; Pei, Y.; Zhang, H.; Liu, Y.; Wang, Q. Dependence on manufacturing directions of tensile behavior and microstructure evolution of selective laser melting manufactured Inconel 625. *J. Mater. Eng. Perform.* **2023**, *32*, 7488–7500. [CrossRef]
27. Hong, M.; Wang, S.; Sun, W.; Geng, Z.; Xin, J.; Ke, L. Effect of welding speed on microstructure and mechanical properties of selective laser melting Inconel 625 alloy laser welded joint. *J. Mater. Res. Technol.* **2022**, *19*, 2093–2103. [CrossRef]
28. Huang, W.; Li, Y.; Yanjie, R.E.N.; Sun, J.; Xia, Z.; Zhou, L.; Li, C.; Chen, J.; Niu, Y.; Zhao, Y. Effect of scanning speed on the high-temperature oxidation resistance and mechanical properties of Inconel 625 alloys fabricated by selective laser melting. *Vacuum* **2022**, *206*, 111447. [CrossRef]
29. Du, K.; Yang, L.; Xu, C.; Wang, B.; Gao, Y. High Strain Rate Yielding of Additive Manufacturing Inconel 625 by Selective Laser Melting. *Materials* **2021**, *14*, 5408. [CrossRef]
30. Zhang, M.; Liang, X.; Zhang, X.; Liu, M.; Chu, Q.; Zhou, X. Effects of Heat Treatment on Microstructure and Properties of Inconel625 Alloy Blades Prepared by Selective Laser Melting. *Int. J. Electrochem. Sci.* **2022**, *17*, 220112. [CrossRef]
31. Pleass, C.; Jothi, S.; Krishnan, M. Grain boundary and triple junction characteristics analytics of additive manufactured Inconel 625 superalloy using selective laser melting. *Mater. Sci. Eng. A* **2023**, *869*, 144744. [CrossRef]
32. Lee, J.; Song, S.; Byeon, M.; Hong, H.U. Suppression of twin formation by bimodal-sized particles and brittle grain boundary fracture in selective laser melted Inconel 625. *Mater. Sci. Eng. A* **2022**, *847*, 143315. [CrossRef]
33. Soni, H.; Gor, M.; Rajput, G.S.; Sahlot, P. A comprehensive review on effect of process parameters and heat treatment on tensile strength of additively manufactured Inconel-625. *Mater. Today Proc.* **2021**, *47*, 4866–4871. [CrossRef]
34. Chaurasia, J.K.; Jinoop, A.N.; Parthasarathy, P.; Paul, C.P.; Bindra, K.S.; Bontha, S. Study of melt pool geometry and solidification microstructure during laser surface melting of Inconel 625 alloy. *Optik* **2021**, *246*, 167766. [CrossRef]
35. Wang, J.; Wang, Y.; Su, Y.; Shi, J. Evaluation of in-situ alloyed Inconel 625 from elemental powders by laser directed energy deposition. *Mater. Sci. Eng. A* **2022**, *830*, 142296. [CrossRef]
36. Hong, K.M.; Grohol, C.M.; Shin, Y.C. Comparative assessment of physics-based computational models on the NIST benchmark study of molten Pool dimensions and microstructure for selective laser melting of Inconel 625. *Integr. Mater. Manuf. Innov.* **2021**, *10*, 58–71. [CrossRef]
37. Sitek, R.; Ciftci, J.; Moszczyńska, D.; Maj, P.; Ura-Bińczyk, E.; Warzybok, P.; Cieślik, I.; Wiśniewski, P.; Mizera, J. Effect of annealing on the microstructure and properties of IN 625 specimens manufactured by selective laser melting. *Arch. Civ. Mech. Eng.* **2022**, *22*, 207. [CrossRef]
38. Hu, Y.; Lin, X.; Li, Y.; Ou, Y.; Gao, X.; Zhang, Q.; Li, W.; Huang, W. Microstructural evolution and anisotropic mechanical properties of Inconel 625 superalloy fabricated by directed energy deposition. *J. Alloys Compd.* **2021**, *870*, 159426. [CrossRef]

39. Hu, Y.; Lin, X.; Li, Y.; Zhang, S.; Zhang, Q.; Chen, W.; Li, W.; Huang, W. Influence of heat treatments on the microstructure and mechanical properties of Inconel 625 fabricated by directed energy deposition. *Mater. Sci. Eng. A* **2021**, *817*, 141309. [CrossRef]
40. Li, B.; Du, J.; Sun, Y.; Zhang, S.; Zhang, Q. On the importance of heat source model determination for numerical modeling of selective laser melting of IN625. *Opt. Laser Technol.* **2023**, *158*, 108806. [CrossRef]
41. Allam, A.; Sugino, C.; Harding, M.; Bishop, D.P.; Erturk, A.; Ruzzene, M. Phased Array Ultrasonic Testing of Inconel 625 Produced by Selective Laser Melting. *J. Nondestruct. Eval. Diagn. Progn. Eng. Syst.* **2021**, *4*, 041006. [CrossRef]
42. Wormald, S.; Clingenpeel, J.; Vincent, T.; Chaudhary, A. Integrated Computational Materials Engineering to Predict Dimensions for Steady-State and Transient Melt-Pool Formation in the Selective Laser Melting of Inconel 625. *Integr. Mater. Manuf. Innov.* **2021**, *10*, 348–359. [CrossRef]
43. Teng, Q.; Li, S.; Wei, Q.; Shi, Y. Investigation on the influence of heat treatment on Inconel 718 fabricated by selective laser melting: Microstructure and high temperature tensile property. *J. Manuf. Process.* **2021**, *61*, 35–45. [CrossRef]
44. Schmeiser, F.; Krohmer, E.; Wagner, C.; Schell, N.; Uhlmann, E.; Reimers, W. In situ microstructure analysis of Inconel 625 during laser powder bed fusion. *J. Mater. Sci.* **2021**, *57*, 9663–9677. [CrossRef]
45. Zhou, L.; Peng, Z.; Chen, J.; Ren, Y.; Niu, Y.; Qiu, W.; Tang, J.; Li, Z.; Li, C. The evolution of microstructure and mechanical properties of Inconel 625 alloy fabricated by laser powder bed fusion via novel hybrid scanning strategy. *Mater. Sci. Eng. A* **2024**, *911*, 146925. [CrossRef]
46. Li, B.; Sun, Y.; Du, J.; Xia, Y.; Su, G.; Zhang, Q. Comparison of Heat Source Model for Numerical Modeling of Selective Laser Melting of IN625 Superalloy. *Trans. Nanjing Univ. Aeronaut. Astronaut.* **2024**, *41*, 174.
47. Moradi, M.; Pourmand, Z.; Hasani, A.; Moghadam, M.K.; Sakhaei, A.H.; Shafiee, M.; Lawrence, J. Direct laser metal deposition (DLMD) additive manufacturing (AM) of Inconel 718 superalloy: Elemental, microstructural and physical properties evaluation. *Optik* **2022**, *259*, 169018. [CrossRef]
48. Angelastro, A.; Campanelli, S.L.; Casalino, G. Statistical analysis and optimization of direct metal laser deposition of 227-F Colmonoy nickel alloy. *Opt. Laser Technol.* **2017**, *94*, 138–145. [CrossRef]

designs

MDPI

Article

Effect of Curvature Shape on the Impact Strength of Additively Manufactured Acrylonitrile Butadiene Styrene Parts Produced via Fused Deposition Modeling

Muhammad Fahad *, Waseem Raja, Muhammad Naveed Iqbal * and Abdul Waheed Awan

Department of Engineering, School of Digital, Technology, Innovation & Business, University of Staffordshire, Stoke-on-Trent ST4 2DE, UK; ww.raja001@gmail.com (W.R.); a.awan@staffs.ac.uk (A.W.A.)
* Correspondence: muhammad.fahad@staffs.ac.uk (M.F.); naveed.iqbal@staffs.ac.uk (M.N.I.)

Abstract: Additive manufacturing (AM) has greatly revolutionized manufacturing due to its ability to manufacture complex shapes without the need for additional tooling. Most AM applications are based on geometries comprising curved shapes subjected to impact loads. The main focus of this study was on investigating the influence of infill density and the radius of curvature on the impact strength of parts manufactured via an FDM process. Standard geometrical specimens with varying part infill densities and radii of curvature were produced and subjected to Charpy impact tests to evaluate their impact strength. The results suggest that the impact strength increases with the increased density caused by higher amounts of material as well as by the changing cross-sectional areas of the beads. Also, the radius of curvature of the parts shows a clear inverse relationship with the impact energy absorbed by the specimens (i.e., increasing the radius decreased the impact energy of the parts) produced via an FDM process, which can be explained using the beam theory of structural mechanics. The maximum value of impact strength obtained was 287 KJ/m^2, and this was achieved at the highest infill density (i.e., solid) and for the smallest radius of curvature.

Keywords: additive manufacturing; 3D printing; fused deposition modeling; acrylonitrile butadiene styrene; Charpy impact testing

1. Introduction

Scientific and technological progress during the twenty-first century has revolutionized modern living as well as manufacturing methods. One such revolution related to manufacturing processes is the advent of additive manufacturing (AM). Additive manufacturing, also known as three-dimensional (3D) printing, is the name given to a group of manufacturing processes which create parts contained in a digital file comprising the computer-aided design (CAD) data of the part [1]. The digital file contains information about the cross-sectional layers of the part at defined heights, and the machine used to produce this part extracts this information to create layers of material that are layered upon each other accordingly to create the geometry of the part. The manufacturing of components and shapes using AM techniques is performed without the need for any additional tooling, resulting in less product development time and fewer costs. At the same time, these processes provide various other advantages such as the consolidation of multiple parts into a single part, a complex design (design freedom), and the production of products with materials of different functional grades. Due to these advantages, three-dimensional printing processes have gained much attention during the last two decades and have shown applications in various industries, including the aerospace, automotive, home appliance, fashion, decoration, and biomedical sectors [1].

According to the ISO and ASTM, seven categories of additive manufacturing processes have been defined [2]. These categories are based on the type of technology used to produce a solid object and include technologies like powder sintering, photo-polymerization, inkjet printing (binder or material), sheet lamination, electron-beam processing, and material extrusion. Material extrusion-based AM processes, commonly known as fused deposition modeling (FDM) or fused filament fabrication (FFF), are the main topics of this paper and will be explained further.

The process of FDM was introduced by Scott Crump [3] during the early 1990s and is based on extruding thin beads of molten thermoplastic material via a heated nozzle to form a layer of material. The nozzle typically moves in both the horizontal and vertical directions via a numerically controlled mechanism. The subsequent addition of layers results in a completed part. The process typically utilizes thermoplastic materials to build parts. Due to its relatively simple technical arrangement, extrusion-based 3D printing category is the most commonly utilized [4,5]. Commonly varied parameters during extrusion-based processes include the following [6–9]:

1. Layer thickness—the height of the deposited layer.
2. Infill density—the mass of the material within the volume of the geometry.
3. Infill patterns—the repetitive pattern of the beads deposited inside a layer.
4. Raster angle—the relative angle at which the beads are deposited within a single layer.
5. The orientation of the part—the relative placement of the part's CAD model within the build chamber.
6. Deposition speed—the relative speed of the extruding nozzle along the XY plane.
7. Build temperature—the temperature at which the material moves out of the nozzle.

Due to their wide variety of materials and their ability to produce complex shapes without needing additional tooling, extrusion-based AM processes have been widely adopted for short production runs (customized production) for a variety of applications [10,11], including automotive components [12,13], biomedical components [14], drug delivery [15], toys [16], energy storage devices [17], and personal protective equipment [18,19]. While extrusion-based AM can utilize numerous materials, acrylonitrile butadiene styrene (ABS) and polylactic acid (PLA) are the most frequently employed and studied materials [8,20–22]. For this research, ABS was chosen due to its superior mechanical properties. Consequently, the subsequent part of this literature review will focus primarily on ABS-related studies and applications.

The effect of varying printing parameters on the mechanical behavior of 3D-printed ABS parts has been discussed by different authors over the previous two decades. The research conducted and published by Fodran et al. [23] is one of the pioneering studies in this field. In their work, Fodran et al. investigated the influence of air gaps, the width, and the raster angle on the Young's modulus, the yield strength, and the tensile strength of parts. Similarly, another comprehensive study was performed by Sood et al. [8], who studied the effect of changing printing parameters on tensile strength. Their research highlighted that the tensile strength of the specimen initially decreased but then increased as the layer thickness increased. From this result, it can be anticipated that with a lower layer thickness, the number of layers will increase, which will result in a greater number of thermal cycles, which further leads to a higher temperature gradient within the previously stacked layers. This will increase the inter-raster and inter-layer bonding, and eventually, the strength of the part will improve. Other similar studies, such as those by Tymrak et al. [24] and Onwubolu et al. [25], indicated similar results. In other studies, Griffith et al. [26], Li et al. [27], Torres et al. [28] and Alafaghani et al. [29] have identified a visible correlation between the tensile strength of ABS parts and the infill density. All these studies suggest that an increase in infill density results in increased tensile strength for the 3D-printed parts. This increment in tensile strength is primarily attributed to the higher amount of material within the same volume. Therefore, parts with higher infill densities can endure higher loads and stresses without deforming or prior to fracturing. Therefore, these results clearly identify infill density as a very important factor in the design of 3D-printed objects for applications where mechanical strength is critical. However, increasing the infill density also increases both the amount of material required to print and the total time required to complete the part. This in turn, increases the total cost of part production at higher densities.

Similarly, the effect of process parameters on compressive strength has been studied comprehensively by many authors [30–35]. Ang K C et al. [30] studied the compressive strength by varying air gaps within 3D-printed parts of ABS and noted an increase in compressive strength upon reducing air gaps. Sood et al. [31] investigated the influence of layer thickness on the compressive behavior of ABS parts and showed that increasing layer thickness results in parts with higher compressive strength. The effect of varying the orientation of parts within the build chamber was studied by Hernandez et al. [32], and it was found that the compressive strength of parts built along the Z-plane was inferior to those built along the XY-plane. Dominguez-Rodríguez et al. [33] also evaluated the effect of build orientation along with different infill patterns on the compressive strength of 3D-printed ABS parts and showed that parts built along the vertical direction have a better compressive strength compared with the parts printed along the transverse direction. Recently, Fahad et al. [34] also studied the influence of process parameters such as part orientation, print density, and layer thickness on compressive strength and indicated that increasing layer thickness and print density have a direct effect on compressive strength.

Subsequent research has extensively explored the impact of process parameters on the mechanical properties of ABS parts. However, the majority of these studies focused primarily on tensile characteristics (such as the strength, elongation, and modulus of elasticity), with comparatively less attention given to impact behavior. Sood et al. [8] investigated the effects of layer thickness and part orientation, revealing that increased layer thickness led to higher impact energy absorption in parts. Notably, they found no significant variation in impact energy when changing part orientation within the XY-plane. Ziemian et al. [35] examined the influence of raster angle on impact behavior, observing a decrease in impact energy as the angle changed from 0° to 90°. Górski et al. [36] further explored part orientation, concluding that parts printed along the X-direction exhibited superior impact energy compared to other orientations. In a similar study, Dawoud et al. [37] investigated the effects of air gap and raster angle, determining that maximum strength was achieved with a raster angle of 45°. These findings collectively contribute to a growing body of knowledge on the relationship between process parameters and the mechanical properties of ABS parts produced through additive manufacturing.

Most of the parts produced via three-dimensional printing processes are complex in shape and comprise curved features. Specific examples of parts with curved shapes and features include helmets, masks, toys, medical implants, and automotive components. The presence of curved features not only influences the accuracy and surface finish but also affects the mechanical properties of the parts [38]. Also, most of the previously mentioned applications are typically subjected to impact loads, as opposed to other types of loads. Therefore, the objective of this study was to examine the influence of infill density and curved shape on the impact behavior of an ABS specimen produced via the fused deposition modeling process. Specimens with three different infill densities and varying radii of curvature were prepared and their impact strength was examined in order to evaluate their suitability for typical applications requiring curved-shaped parts printed via the FDM process.

2. Materials and Methods

2.1. Material

The material used to produce the test specimen in the study was Acrylonitrile Butadiene Styrene (ABS) plus material (filament diameter = 1.75 mm) manufactured by Stratasys (Edina, MN, USA). Important properties of the used material are indicated in Table 1. The material was used as received by the supplier without any further modifications.

Table 1. Printing parameters used in the research.

Property	Value	Unit
Yield Strength	31	MPa
Ultimate Tensile Strength	33	MPa
Elongation at Break	6%	-
Glass Transition Temperature	108	°C
Coefficient of thermal Expansion	8.82×10^{-5}	$°C^{-1}$
Specific Gravity	1.04	-

2.2. Parts Printing

A uPrint SE FDM machine, manufactured and purchased by Stratasys (USA), was used to print the specimens. Three different infill densities, namely low, high, and solid, were used. The values of infill percentage were 53% at low, 81% at high, and 97% at solid. Three replicates of each type of specimen were produced, and the average results for each type were used to make a comparison. The parameters used to print the parts are indicated in Table 2. Parts were oriented during printing in the build chamber such that the length (L) was long on the x-axis, the width (W) was along y, the y-axis, and the thickness (T) was along z, the z-axis (Figure 1). The actual printed specimens are shown in Figure 2.

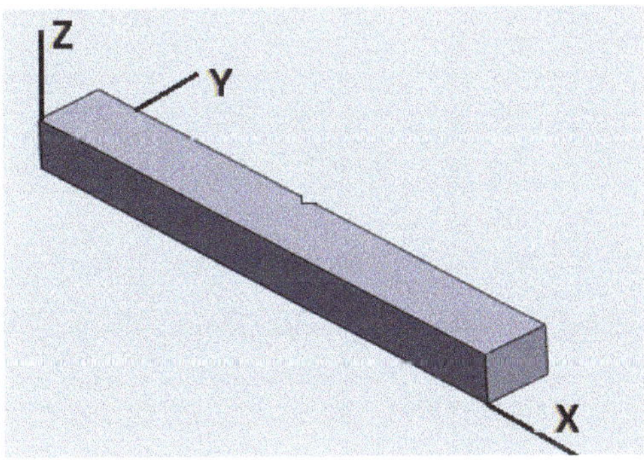

Figure 1. Part orientation within the build chamber during printing.

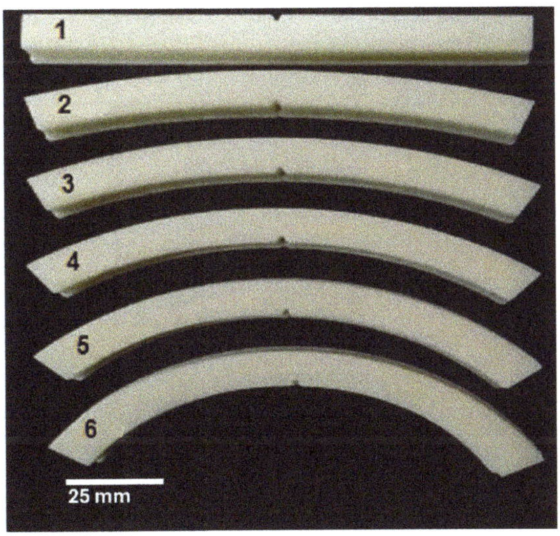

Figure 2. Six different types of specimens were printed for research.

Table 2. Printing parameters used in the research.

Parameter	Value	Unit
Nozzle Temperature	300	°C
Build Envelop Temperature	80	°C
Layer Thickness	0.254	mm
Nozzle diameter	0.4	mm
Raster Orientation	+45/−45	degrees

2.3. Impact Testing

Standard straight specimens for Charpy impact testing were produced according to ASTM D6110 [39] (Figure 3), where the length (L) was 127 mm, and the width (W) and the thickness (T) were 12.7 mm each. This straight sample was converted into a curved shape by changing the height (h) or radius of curvature (R) of a segment of the circle, as shown in Figure 3b. The cross-sectional area and the length of each specimen were kept constant to ensure that the overall volume remained constant. Six different specimen shapes (Figure 2), including one straight and five different curved shapes with a height (h) varying from 10 to 30 mm in increments of 5 mm, were used for building the specimen (Table 3). The impact strength was measured using the Charpy impact test. A pendulum impact energy tester G.U.N.T model WP-410 was used for the impact testing (Figure 4). The impact speed of the hammer was 5.5 m/s, and the total capacity of the hammer was 300 Nm. The impact energy was calculated by measuring the distance (d) traveled by the hammer after impact using the following equation, where m = mass of hammer, g = acceleration due to gravity, h = height of hammer, d = distance traveled by a hammer after impact and A = cross-sectional area of the specimen.

$$\text{Energy} = \text{Impact Strength} = mg\,(h - d)/A$$

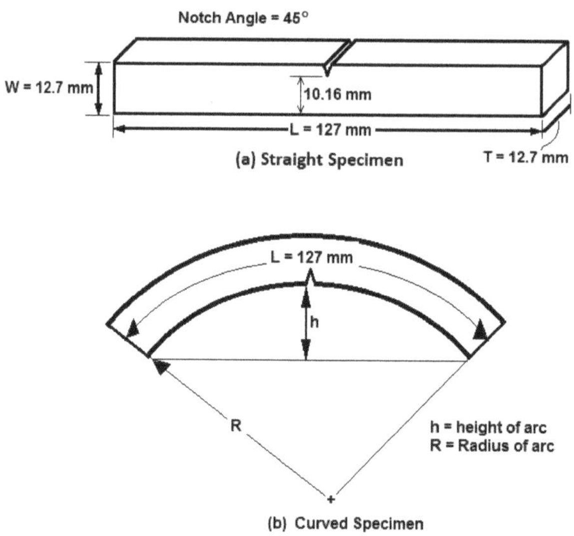

Figure 3. Dimensions of the impact specimens used in the research: (**a**) straight specimen; (**b**) curved specimen.

Table 3. Values of radius (R) and height (h) of the arc used in the research.

Part No.	Length of Arc (mm), L	Radius of Arc (mm), R	Height of Arc (mm), h
1	127	∞	0
2	127	199.92	10
3	127	131.82	15
4	127	97.27	20
5	127	76.04	25
6	127	61.42	30

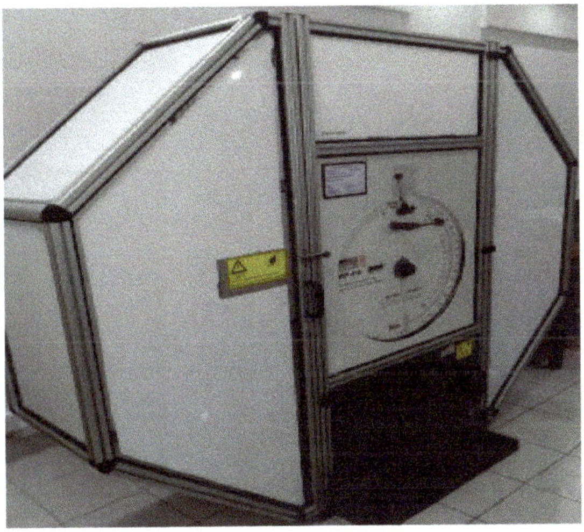

Figure 4. Impact testing machine used in the research (GUNT WP410).

2.4. Scanning Electron Microscopy (SEM)

In order to analyze the cross-section of the broken samples after impact testing, Scanning Electron Microscopy (SEM) was used. Images were taken using analytical SEM by JEOL (Model Number: JSM-6380). The samples were sputtered with thin layers of gold using JEOL JFC-1500 sputtering equipment.

3. Results

As mentioned previously, six different samples were printed, with varying radii and heights of curvature (three replicates of each), as shown in Figure 2. The impact strength results were obtained by calculating the average of three samples of each type. The average values of each sample type are indicated in Table 4. Variations in impact strength with changes in infill density for different specimens are indicated in Figure 5, whereas Figure 6 shows the variations in impact strength with changes in the radius of curvature of specimens. Images of broken samples are shown in Figure 7.

Table 4. Average impact strength values for different specimens.

Infill Density → (g/cm^3)	Low 0.535				High 0.81				Solid 0.97			
Part No. ↓	Impact Strength (KJ/m^2)											
	1	2	3	Avg.	1	2	3	Avg.	1	2	3	Avg.
1	31.14	31.6	31.46	**31.4**	52.39	53.62	52.6	**52.9**	118.32	119.1	117.78	**118.4**
2	35.04	34.18	34.35	**34.5**	87.01	85.8	85.23	**86.0**	119.01	119.73	118.61	**119.1**

Table 4. *Cont.*

Infill Density → (g/cm³)	Low 0.535				High 0.81				Solid 0.97			
Part No. ↓	**Impact Strength (KJ/m²)**											
	1	**2**	**3**	**Avg.**	**1**	**2**	**3**	**Avg.**	**1**	**2**	**3**	**Avg.**
3	35.2	34.85	34.96	**35.0**	88.23	88.08	87.29	**87.9**	222.97	226.07	225.29	**224.8**
4	38.7	38.97	38.93	**38.9**	113.89	114.98	115.43	**114.8**	283	280.68	278.95	**280.9**
5	42.04	41.81	41.88	**41.9**	154.93	152.04	152.91	**153.3**	286.6	284.55	283.97	**285.0**
6	125.64	124.12	123.2	**124.3**	254.32	256.57	256.12	**255.7**	287.8	286.41	286.88	**287.0**

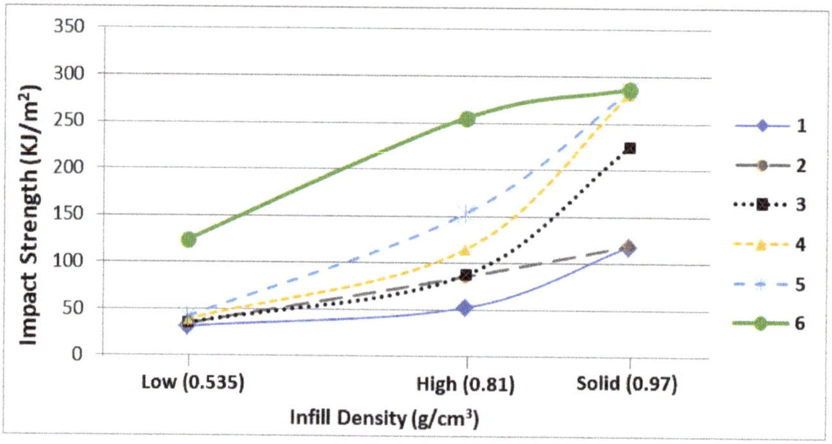

Figure 5. Variation in impact strength with print density.

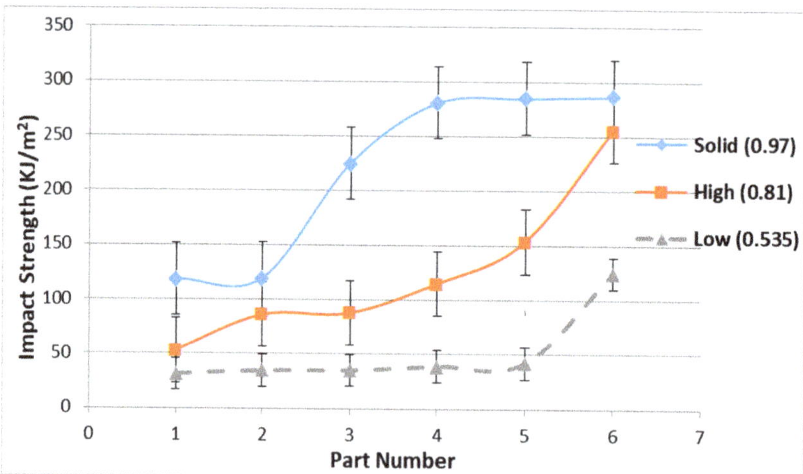

Figure 6. Variation in impact strength with part curvature.

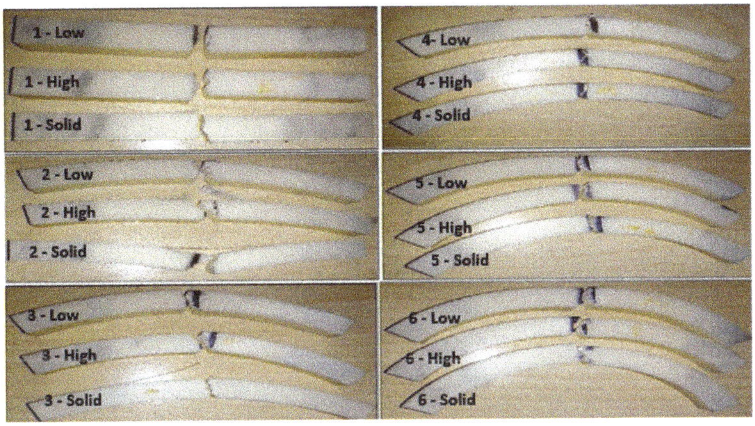

Figure 7. Images of samples after impact testing.

4. Discussion

4.1. Effect of Infill Density

Unlike most available open-source 3D printers, the 3D printer used in this research did not allow the print density to be set at any quantitative value. Rather, three options, low, high, and solid, could be selected to build parts. Therefore, to find the actual value of infill density, the mass of parts printed at different densities was measured and divided by the volume (constant for all parts) to obtain the actual density values for the three levels. These values were found to be 0.535 g/cm^3, 0.81 g/cm^3, and 0.97 g/cm3 for low, high, and solid densities, respectively. The variation in impact strength with print density is shown in Figure 5, and it is evident that the impact strength increased with the increase in print density. It was shown previously that increasing print density results in increased tensile strength and impact strength [34,37,40]. This can be attributed to the close packing of material inside the same volume at higher print densities, resulting in more energy being required for fracture. A scanning electron microscopy (SEM) of the fractured specimen was conducted, and the results are indicated in Figures 8 and 9. Both Figures 8 and 9 clearly indicate that the material failed under the brittle fracture mode, and this is consistent with the material behavior defined by the manufacturer (only 6% elongation at break) in Table 1. It is clear from Figures 8a and 9a that for the lowest-density sample, more empty spaces are present, and these void spaces act as pre-existing cracks to initiate fracture. On the other hand, for the high-density (Figures 8b and 9b) and solid-density samples (Figures 8c and 9c), the internal structure is much more compact and contains very few empty spaces. The decrease in the presence of crack-initiating points results in a higher impact energy requirement for the fracture of parts with increased infill density. It is also evident from Figures 8 and 9 that the shape of the beads in lower-density samples is consistently circular. For the highest-density (i.e., solid) samples, the circular bead shape of changes to an elliptic shape. It is well known in mechanics that, for the same cross-sectional area, the elliptic shape has a higher moment of inertia along the major axis (the axis along the width of the bead) compared to the moment of inertia of the circular cross-section [41,42]. This higher moment of inertia also strengthens the internal structure and results in higher energy absorption before fracture as the density of the specimen is increased. It may be noted that the high amount of material that is present mainly causes this change in the cross-sectional shape of beads in higher-density specimens with the same volume (i.e., a high infill density). The excess weight presses the previously deposited beads into an elliptical shape. On the other hand, for low-density samples, the amount of material on top of the beads is not sufficiently large to cause this elliptic deformation.

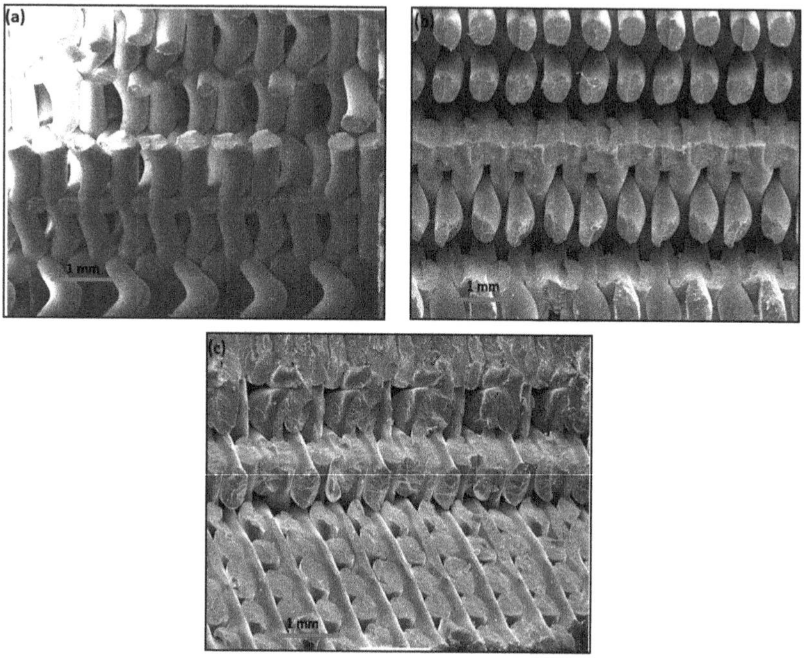

Figure 8. SEM images of part 1 (0 mm height of arc) at: (**a**) low density; (**b**) high density; (**c**) solid density.

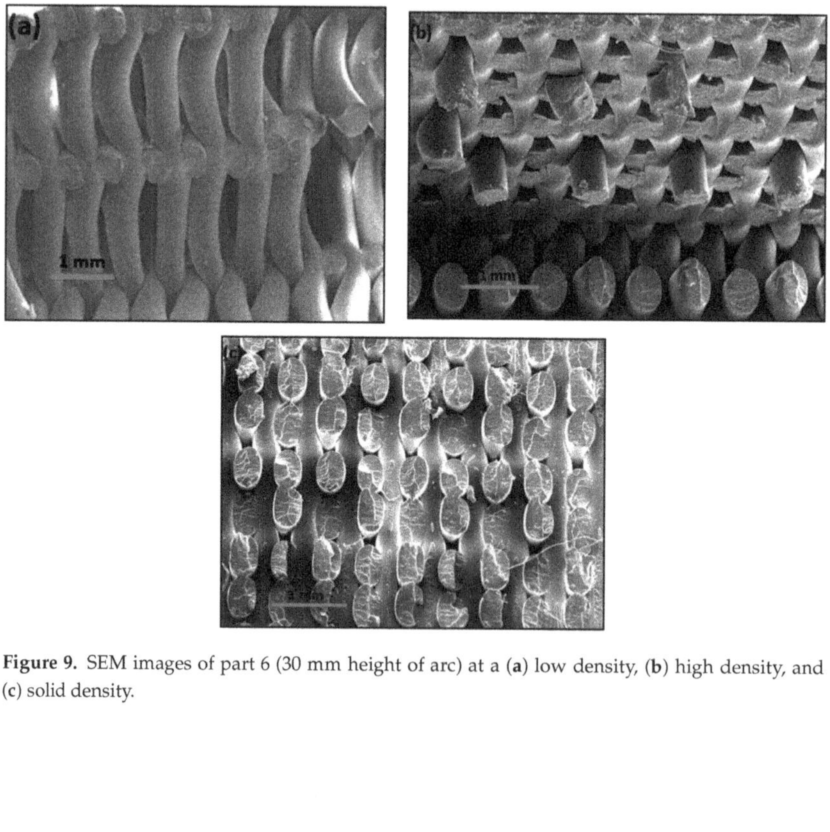

Figure 9. SEM images of part 6 (30 mm height of arc) at a (**a**) low density, (**b**) high density, and (**c**) solid density.

4.2. Effect of Curvature

The trend of impact strength changing with the changing radius of curvature at each infill density is presented in Figure 6, and it is clear that decreasing the radius of curvature (indicated by increasing part number) at a fixed infill density results in increased impact energy for samples of all densities. Although no study related to curved 3D-printed structures is available, curved composite structures have been subjected to impact loads, and similar results were achieved [43,44]. The reason behind this increased impact strength can be explained using the commonly used bending stress equation for curved beams (Figure 10), as presented in Equation (1), where σ_b = bending stress; E = Young's modulus; y = distance (from neutral axis) of the layer subjected to bending; R = radius of curvature:

$$\sigma_b = \frac{E.y}{R}, \quad (1)$$

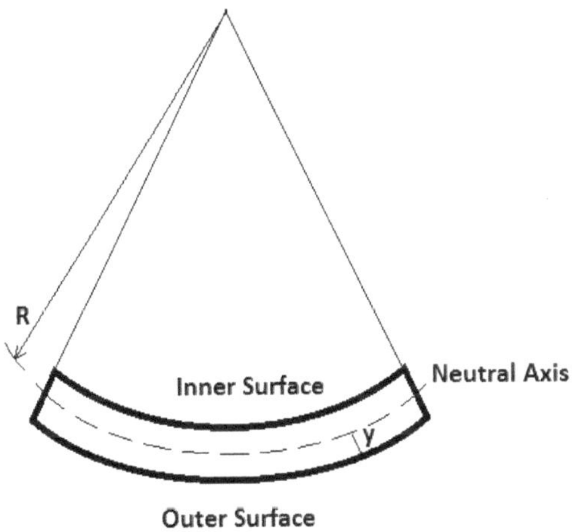

Figure 10. Schematic diagram of a curved beam.

Curved beams have an internal moment or bending moment, which is composed of tensile and compressive stresses at their outer surfaces, as given by Equation (1). This means that the outer surface of the curved specimen is under tensile stress, and the inner side is under compressive stress. When the impact load is applied on the outer surface of the curved specimen, the presence of tensile stress results in an increase in the amount of energy needed to fracture the specimen. As the radius of the arc (i.e., R) is decreased from samples 1 to 6, the bending stresses increase, and therefore, the energy required to break the specimen increases from samples 1 to 6.

However, it is clear from Figure 6 that a sharp rise in impact energy occurred in the highest-density samples at relatively higher radius values (i.e., a sharp rise between parts 2 and 3). This indicates that for higher-density samples, the impact strength for a relatively low depth of curvature is also high. This may be attributed to the fact that at a solid density, the beads within the cross-section are very tightly packed. However, increasing the curvature (reduced radius of curvature) causes the voids, which act as crack initiation points, to come closer; therefore, a sharp rise in impact strength takes place between parts 3 and 4. Therefore, printing parts at a higher density can provide good a impact strength even at relatively high or arc values. An appreciable rise in impact strength for low-density samples was obtained at a low arc radius (increased depth). This means when printing parts at a low density, to obtain a suitable impact strength, a smaller radius of curvature would be preferred. This may be because, at low densities, the voids acting as crack initiation points move closer at a small radius of curvature, as indicated by Figures 8a and 9a. On the other hand, the samples printed with medium density values (i.e., 0.81

g/cm^3) showed a nearly linear behavior of increasing impact strength at reduced radius values for the arc. This may be because the compactness of the voids remains consistent at all curvatures for samples with a density between low and solid. This implies that a medium-print density is more suited for the entire arc radius range.

These results show that the use of the FDM process for producing end-use items containing curved shapes (helmets, masks, medical implants, etc.) is beneficial in the sense that these parts have increased strength and are able to withstand higher impact loads than parts with flat shapes. The results of this study can help users of commercial FDM processes to produce curved parts by selecting appropriate curve radii values within their part design and identifying the appropriate infill density. Selecting the right curve radius will ensure that the part is produced with the desired impact strength, making it suitable for its end application. Additionally, the selection of an appropriate infill density will help in the manufacturing of parts by ensuring an appropriate amount of material and a appropriate printing time are used. The printing time and the amount of material required to print parts are directly linked to the cost and overall environmental impact of additive manufacturing. Therefore, these results provide users with suitable values for the infill density and suitable curvature values to ensure that the target strength is achieved while ensuring minimal costs and a minimal environmental impact. These results can help when using the FDM process to build parts for applications that require curved-shape components. These applications include toy manufacturing, automotive parts, biomedical components, jewelry items, and home appliances.

5. Conclusions

Specimens with varying print densities and curvature were manufactured using ABS material using a fused deposition modeling process. The printed specimens were subjected to Charpy impact testing to evaluate the influence of infill density and the radius of curvature on the impact strength. The main findings of the study are as follows:

- The results indicate that the impact strength increased upon increasing the infill density for all parts due to the increased amount of material inside an identical volume. This is consistent with the literature. The main reason behind this increase is the close packing of material inside the cross-section.
- SEM results indicated a change in bead shape from circular to elliptical upon increasing infill density. This change in bead shape also helps increase energy absorption before fracture.
- The results further showed that decreasing the arc radius resulted in an increased impact strength of the parts at all densities.

This increased impact strength shows that the FDM technique is suitable for producing different curved-shaped end-use items such as helmets, toys, masks, and automotive components. The results can help users of the FDM process in selecting an appropriate combination of infill density and curve radii to ensure the correct amount of material and print time are selected for printing parts while achieving the desired impact strength.

Author Contributions: Conceptualization, M.F. and W.R.; methodology, M.F., M.N.I. and W.R.; software, W.R., M.N.I. and A.W.A.; validation, M.F. and A.W.A.; formal analysis, A.W.A., M.N.I. and W.R.; writing—original draft preparation, M.F. and W.R.; writing—review and editing, M.N.I. and A.W.A.; visualization, A.W.A. and M.N.I.; supervision, M.F. All authors have read and agreed to the published version of the manuscript.

Funding: This research received no external funding.

Data Availability Statement: Data are contained within the article.

Acknowledgments: Authors acknowledge Karachi University for providing support during scanning electron microscopy of the samples.

Conflicts of Interest: The authors declare no conflicts of interest.

References

1. Fahad, M. Additive Manufacturing. In *Foundations of Materials Science and Engineering*; Trans Tech Publications: Stafa-Zurich, Switzerland, 2017; Volume 93, pp. 355–376.
2. *ISO/ASTM 52900*; Additive Manufacturing—General Principles—Terminology. International Organization for Standardization: Geneva, Switzerland, 2015.
3. Crump, S.S. Apparatus and Method for Creating Three-Dimensional Objects. U.S. Patent US 5121329 A, 9 June 1992.
4. Tessa, J.G.; Philipp, R.T.; Louis, T.; Lars, J. Optimising the FDM additive manufacturing process to achieve maximum tensile strength: A state-of-the-art review. *Rapid Prototyp. J.* **2019**, *25*, 953–971.
5. Andrei, D.M.; Dumitru, N.; Ramona, P. Additive manufacturing of composite materials by FDM technology: A review. *Indian J. Eng. Mater. Sci.* **2019**, *27*, 179–192.
6. Popescu, D.; Zapciu, A.; Amza, C.; Baciu, F.; Marinescu, R. FDM process parameters influence over the mechanical properties of polymer specimens: A review. *Polym. Test.* **2018**, *69*, 157–166. [CrossRef]
7. Montero, M.; Roundy, S.; Odell, D.; Ahn, S.H.; Wright, P.K. Material characterization of fused deposition modeling (FDM) ABS by designed experiments. *Soc. Manuf. Eng.* **2001**, *10*, 1–21.
8. Sood, A.K.; Ohdar, R.K.; Mahapatra, S.S. Parametric appraisal of mechanical property of fused deposition modelling processed parts. *Mater. Des.* **2010**, *31*, 287–295. [CrossRef]
9. Madan, M.N.; Harshavardhan, G.K.; Biradar, D.M. An Overview: Fused Deposition Modeling Process. *J. Emerg. Technol. Innov. Res.* **2020**, *7*, 2115–2119.
10. Whlean, C.; Sheahan, C. Using additive manufacturing to produce injection moulds suitable for short series production. *Procedia Manuf.* **2019**, *38*, 60–68. [CrossRef]
11. Roberson, D.; Shemelya, C.M.; MacDonald, E.; Wicker, R. Expanding the applicability of FDM-type technologies through materials development. *Rapid Prototyp. J.* **2015**, *21*, 137–143. [CrossRef]
12. Romero, P.E.; Arribas-Barrios, J.; Rodriguez-Alabanda, O.; González-Merino, R.; Guerrero-Vaca, G. Manufacture of polyurethane foam parts for automotive industry using FDM 3D printed molds. *CIRP J. Manuf. Sci. Technol.* **2021**, *32*, 396–404. [CrossRef]
13. Maurya, N.K.; Rastogi, V.; Singh, P. Comparative study and measurement of form errors for the component printed by FDM and PolyJet process. *Instrum. Mes. Métrologie* **2019**, *18*, 353–359. [CrossRef]
14. Singh, D.; Singh, R.; Boparai, K.S. Development and surface improvement of FDM pattern based investment casting of biomedical implants: A state of art review. *J. Manuf. Process.* **2018**, *31*, 80–95. [CrossRef]
15. Linares, V.; Casas, M.; Caraballo, I. Printfills: 3D printed systems combining fused deposition modeling and injection volume filling. Application to colon-specific drug delivery. *Eur. J. Pharm. Biopharm.* **2019**, *134*, 138–143. [CrossRef] [PubMed]
16. León-Cabezas, M.A.; Martínez-García, A.; Varela-Gandía, F.J. Innovative functionalized monofilaments for 3D printing using fused deposition modeling for the toy industry. *Procedia Manuf.* **2017**, *13*, 738–745. [CrossRef]
17. Chang, P.; Mei, H.; Zhou, S.; Dassios, K.G.; Cheng, L. 3D printed electrochemical energy storage devices. *J. Mater. Chem. A* **2019**, *7*, 4230–4258. [CrossRef]
18. Wang, P.; Yang, J.; Hu, Y.; Huo, J.; Feng, X. Innovative design of a helmet based on reverse engineering and 3D printing. *Alex. Eng. J.* **2021**, *60*, 3445–3453. [CrossRef]
19. Jafferson, J.M.; Pattanashetti, S. Use of 3D printing in production of personal protective equipment (PPE)—A review. *Mater. Today Proc.* **2021**, *46*, 1247–1260. [CrossRef]
20. Novakova-Marcincinova, L.; Novak-Marcincin, J. Testing of materials for rapid prototyping fused deposition modelling technology. *Int. J. Ind. Manuf. Eng.* **2012**, *6*, 2082–2085.
21. Novakova-Marcincinova, L.; Novak-Marcincin, J. Applications of rapid prototyping fused deposition modeling materials. In *Annals of DAAAM for 2012 & Proceedings of the 23rd International DAAAM Symposium 2012*; AAAM International: Vienna, Austria, 2012; Volume 23, pp. 57–60.
22. Lay, M.; Thajudin, N.L.; Hamid, Z.A.; Rusli, A.; Abdullah, M.K.; Shuib, R.K. Comparison of physical and mechanical properties of PLA, ABS and nylon 6 fabricated using fused deposition modeling and injection molding. *Compos. Part B Eng.* **2019**, *176*, 107341. [CrossRef]
23. Fodran, E.; Koch, M.; Menon, U. Mechanical and dimensional characteristics of fused deposition modeling build styles. In Proceedings of the Solid Freeform Fabrication Proceedings, Austin, TX, USA, 12–14 August 1996; pp. 419–442.
24. Tymrak, B.M.; Kreiger, M.; Pearce, J.M. Mechanical properties of components fabricated with open-source 3-D printers under realistic environmental conditions. *Mater. Des.* **2014**, *58*, 242–246. [CrossRef]
25. Onwubolu, G.C.; Rayegani, F. Characterization and Optimization of Mechanical Properties of ABS Parts Manufactured by the Fused Deposition Modeling Process. *Int. J. Manuf. Eng.* **2014**, *2014*, 598531. [CrossRef]
26. Griffiths, C.A.; Howarth, J.; Rowbotham, G.d.-A.; Rees, A. Effect of Build Parameters on Processing Efficiency and Material Performance in Fused Deposition Modelling. *Procedia CIRP* **2016**, *49*, 28–32. [CrossRef]
27. Li, H.; Wang, T.; Sun, J.; Yu, Z. The effect of process parameters in fused deposition modelling on bonding degree and mechanical properties. *Rapid Prototyp. J.* **2018**, *24*, 80–92. [CrossRef]
28. Torres, J.; Cole, M.; Owji, A.; DeMastry, Z.; Gordon, A.P. An approach for mechanical property optimization of fused deposition modeling with polylactic acid via design of experiments. *Rapid Prototyp. J.* **2016**, *22*, 387–404. [CrossRef]

29. Alafaghani, A.; Qattawi, A.; Alrawi, B.; Guzman, A. Experimental Optimization of Fused Deposition Modelling Processing Parameters: A Design-for-Manufacturing Approach. *Procedia Manuf.* **2017**, *10*, 791–803. [CrossRef]
30. Ang, K.C.; Leong, K.F.; Chua, C.K. Chandrasekaran M. Investigation of the mechanical properties and porosity relationships in fused deposition modeling-fabricated porous structures. *Rapid Prototyp. J.* **2006**, *12*, 100–105.
31. Sood, A.K.; Ohdar, R.K.; Mahapatra, S.S. Experimental investigation and empirical modeling of FDM process for compressive strength improvement. *J. Adv. Res.* **2011**, *3*, 81–90. [CrossRef]
32. Hernandez, R.; Slaughter, D.; Whaley, D.; Tate, J.; Asiabanpour, B. Analyzing the tensile, compressive, and flexural properties of 3D printed ABS P430 plastic based on printing orientation using fused deposition modeling. In Proceedings of the Solid Freeform Fabrication 2016: Proceedings of the 26th Annual International Solid Freeform Fabrication Symposium, Austin, TX, USA, 8–10 August 2016; pp. 939–950.
33. Domínguez-Rodríguez, G.; Ku-Herrera, J.J.; Hernández-Pérez, A. An assessment of the effect of printing orientation, density, and filler pattern on the compressive performance of 3D printed ABS structures by fuse deposition. *Int. J. Adv. Manuf. Technol.* **2018**, *95*, 1685–1695. [CrossRef]
34. Fahad, M.; Mujeeb, M.; Khan, M.A. Effect of Process Parameters on the Compressive and Impact Strength of 3D Printed Parts. *Iran. J. Sci. Technol. Trans. Mech. Eng.* **2022**, *47*, 257–265. [CrossRef]
35. Ziemian, C.; Sharma, M.; Ziemian, S. Anisotropic mechanical properties of ABS parts fabricated by fused deposition modelling. *Mech. Eng.* **2012**, *23*, 159–180.
36. Górski, F.; Wichniarek, R.; Kuczko, W.; Zawadzki, P.; Buń, P. Strength of ABS parts produced by Fused Deposition Modelling technology–a critical orientation problem. *Adv. Sci. Technol. Res. J.* **2015**, *9*, 12–19. [CrossRef]
37. Dawoud, M.; Taha, I.; Ebeid, S.J. Mechanical behaviour of ABS: An experimental study using FDM and injection moulding techniques. *J. Manuf. Process.* **2016**, *21*, 39–45. [CrossRef]
38. Huang, B.; Singamneni, S. Alternate slicing and deposition strategies for fused deposition modelling of light curved parts. *J. Achiev. Mater. Manuf. Eng.* **2012**, *55*, 511–517.
39. Fernandez-Vicente, M.; Calle, W.; Ferrandiz, S.; Conejero, A. Effect of infill parameters on tensile mechanical behavior in desktop 3D printing. *3D Print. Addit. Manuf.* **2016**, *3*, 183–192. [CrossRef]
40. Rodríguez-Panes, A.; Claver, J.; Camacho, A.M. The influence of manufacturing parameters on the mechanical behaviour of PLA and ABS pieces manufactured by FDM: A comparative analysis. *Materials* **2018**, *11*, 1333. [CrossRef]
41. Moment of Inertia—Elliptical Beam. Available online: https://www.physicsforums.com/threads/moment-of-inertia-elliptical-beam.244049 (accessed on 15 September 2024).
42. Pope, J.E. *Rules of Thumb for Mechanical Engineers*; Gulf Pub. Co.: Houston, TX, USA, 1997.
43. Saghafi, H.; Minak, G.; Zucchelli, A. Effect of preload on the impact response of curved composite panels. *Compos. Part B Eng.* **2014**, *60*, 74–81. [CrossRef]
44. Baba, B.O. Curved sandwich composites with layer-wise graded cores under impact loads. *Compos. Struct.* **2017**, *159*, 1–11. [CrossRef]

designs

MDPI

Article

Mechanical Properties of 17-4 PH Stainless Steel Manufactured by Atomic Diffusion Additive Manufacturing

Animesh Kumar Basak [1,*], Jasim Mohammed Sali [2] and Alokesh Pramanik [2,*]

[1] Adelaide Microscopy, The University of Adelaide, Adelaide, SA 5005, Australia
[2] School of Civil and Mechanical Engineering, Curtin University, Bentley, WA 6102, Australia; jasim.sali@postgraduate.curtin.edu.au
* Correspondence: animesh.basak@adelaide.edu.au (A.K.B.); alokesh.pramanik@curtin.edu.au (A.P.)

Abstract: Atomic diffusion additive manufacturing (ADAM) is a specialized extrusion-based metal additive manufacturing (MAM) process where metal parts are produced through a three-stage process of printing, de-binding and sintering. Several scientific facts, such as dimensional error, surface quality, tensile behavior and the internal structure of this process for specific materials for certain conditions, are not well explained in the existing literature. To address these issues, the present manuscript investigates the effect of infill type and shell thickness on 17-4 precipitation-hardened (PH) stainless steels on the dimensional accuracy, surface roughness and mechanical properties of the printed specimens. It was found that the strength (maximum ultimate tensile strength up to 1049.1 MPa) and hardness (290 HRB) of the specimens mainly depend on shell thickness, while infill type plays a relatively minor role. The principle of atomic diffusion may be the reason behind this pattern, as an increase in shell thickness is essentially an increase in the density of material deposited during printing, allowing more fusion during sintering and thus increasing its strength. The two different infill types (triangular and gyroid) contribute towards minimal changes, although it should be noted that triangular specimens exhibited greater ultimate tensile strength, whereas the gyroid had slightly longer elongation at break. Dimensional accuracy and surface roughness for all the specimens remain reasonably consistent. The cross-section of the tensile tested specimens revealed significant pores in the microstructure that could contribute to a reduction in the mechanical properties of the specimens.

Keywords: additive manufacturing; tensile strength; hardness; surface roughness; dimensional accuracy

1. Introduction

Stainless steel (SS) is one of the most important structural materials used in different sectors like industrial machineries, automotive, household and so on. Stainless steel has immense tensile (up to 579 MPa for SS316L) and yield strength (up to 310 MPa for SS316L), as well as ductility (up to 70%) [1]. In many industries, 17-4 precipitation-hardened (PH) stainless steel is utilized for its corrosion resistance and high mechanical strength. Though its wear resistance is lower than that of hard coatings [2], its combined corrosion–wear resistance [3] is superior. Applications of 17–4 PH stainless steel include those in the marine, chemical and power-generating sectors [4]. From start to finish, the manufacturing of steels consists of various techniques, starting from ingot-making by casting. After that, various subtractive manufacturing processes are employed that lead to steel wastage (saw dust, chips and off-cuts), increasing the final product costs. This calls for additional steps to clean up and dispose of the waste appropriately. In that respect, 'additive manufacturing (AM)' opens up new avenues for the production of net/near-net-shaped components, which not only minimizes such waste of materials but also delivers high-quality parts [5].

Atomic diffusion additive manufacturing (ADAM) [6] is a three-stage extrusion-based metal additive manufacturing (MAM) [7] process with the traits of higher accuracy, complicated geometry formation, design flexibility, material savings and personalization while keeping product costs to a low level. In contrast to traditional subtractive manufacturing (e.g., cutting, drilling, grinding and machining), this process involves iteratively adding materials layer upon layer [8]. Extrusion-based MAM techniques are viable alternatives to subtractive manufacturing, since the equipment is less expensive, and no loose powders or lasers are utilized during the production process [9]. The formation of a powder–binder blend is the first step in this manufacturing process. The next step is the printing, where the mixture is subsequently pushed through a nozzle, which relies on the Fused Deposition Modelling (FDM) concept to deposit material layer by layer and form a 3D object. To create the dense metallic components, debinding and sintering are required in the post-printing process [10]. Dense, additively manufactured metal parts offer equal/better quality than that of conventionally manufactured counterparts by controlling their porosity and microstructure [11]. Crack propagation can be mitigated by limiting the porosity of the material, which can be accomplished by controlling the applied energy volume and quality of the feedstock [12]. The finer microstructure in additively manufactured materials generally increase their ultimate and yield strengths [13]. The build orientation also has an impact on the strength of the material, where a tensile stress in one direction will be able to resist a higher load than stress in the other direction [14] and thus give rise to anisotropy in the as-built components [15].

In the experiment carried out by Gabilondo et al. [10], the influence of specimen print orientation was explored. Through tensile testing, it was found that the horizontally printed specimens were much stronger (530 MPa) than the vertically printed specimens (370 MPa). Under examination, it was also found that the porosity of the horizontal specimens (5.67%) was lower than that of the vertical specimens (6.56%). It was also established that the horizontal samples had lines along the axial direction, causing any cracks that propagated to reach the end more easily. Another study conducted by Chacón et al. [16] showed the same trend, where 3D-printed samples show anisotropic behavior. Lower mechanical properties of the samples were observed when printed upright, whereas flat/horizontal samples had much higher strength and stiffness. A build orientation parallel to infill lines will stop the cracks from propagating, whereas in perpendicular printed specimens, infill lines will easily allow cracks to propagate. Burgess et al. [17] found great variation between specimens printed vertically and horizontally in terms of mechanical properties. A MarkForged metal X^{TM} unit was employed to print the specimens on varying axis and angles, as well as a vertical free-standing specimen. The staircase effect [18] was dominant in specimens printed at an angle other than 0° or 90°. Theoretically, the sintering process should eliminate any reason for variation in mechanical properties for specimens printed at different orientations, as the fusion occurs between metal stands in the furnace. Strength should be increased regardless of specimen print orientation; however, this was not the case, as reported in the literature [19]. Henry et al. [1] also came to a similar conclusion after testing a MarkForged Metal X printed specimen, printing at varying print directions. Samples printed with the shell layers oriented perpendicular to the loading direction during tensile testing resulted in poor mechanical properties compared to samples printed with shell layers oriented parallel to the loading direction. Sambrook et al. [20] carried out experiments on 17-4 PH stainless steel printed by a MarkForged Metal X printer to print with horizontal and vertical print directions. According to their reports, horizontally printed specimens performed 14.7% better in Young's modulus, 64% greater in yield strength and 84.5% greater in ultimate tensile strength, with an increase of 67.1% in elongation at break when compared to vertically printed specimens under the same printing parameters.

Rosnitschek et al. [21] conducted a study on 316L stainless steel manufactured additively by a Desktop Metal Studio printer. The focus was on infill degree and a hexagonal infill pattern with four different infill densities. It was found that 75% and 50% infill patterns provided the strongest specimens, while the 25% infill specimen performed worse than those two in terms of Young's modulus and tensile stress. It was surprising to note that the 100% infill specimen performed worse than the 50% and 75% specimens. Rosnitschek et al. [21] believed that this occurred due to the presence of printing flaws, as well as thermal stress build-up owing to the deposition of a large quantity of material, which hampered the components' homogeneous cooling. Thawon et al. [22] tensile tested 316L and 17-4 PH stainless steel at different infill densities to determine their mechanical behavior. The testing found the 316L specimens to be of ductile nature, whereas the 17-4 PH specimens were of brittle nature, with higher yield strength both in the elastic region and over the whole strain dimension. In addition, there was a significant increase in the mechanical properties of the material with an increase in infill densities. The non-solid specimens had a linear increase in mechanical properties, whereas the solid specimen (100% infill) had an exponential increase in mechanical properties. Gabilondo et al. [10] investigated the hardness of 316L stainless steel specimens fabricated at varying infill patterns and orientations by the Bound Metal Deposition (BMD) [23] process. They reported an average hardness of 120 HV, which was significantly lower than that of the same specimens fabricated by either Powder Bed Fusion (PBF), 230–280 HV or traditional casting, 165 HV. This was due to the existence of pores, voids and intergranular porosity in the BMD specimens. The low cooling during sintering was a suspected reason that may have contributed to that.

Godec et al. [24] investigated the layer thickness, extrusion temperature and flow rate multiplier of 17-4 PH stainless steel manufactured by Fused Filament Fabrication (FFF). It was found that the layer thickness had a direct correlation with the tensile strength. A larger layer thickness requires fewer layers for the part to be printed; this, in turn, allows for less area for voids to form and creates an area of weak spots in the specimen. The decrease in the interlayer area will allow for a stronger tensile output, as less of the part's strength is reliant on the fusion between layers. Similar results were also reported by Shakeri et al. [25], where a large layer height results in more deformation and geometrical inaccuracies. There are many more investigations on the properties of atomic diffusion additive manufactured 17-4 PH. Raju et al. [26] investigated the printability of holes at different orientations and explored the properties of the green samples with SEM and CT scan. Rodriguez et al. [27] investigated the effect of printing orientation on density, surface roughness, hardness, tensile properties and fatigue behavior. Bouaziz et al. [28] explored the effect of layer thickness on microstructural pattern during tensile testing. They [29] also investigated the effect of heat treatment on the surface roughness and tensile strength.

Because it is a relatively new technique, it is clear from above discussion that there is no investigation on the effect of wide range printing parameters such as shell thickness and infill type on dimensional accuracy, roughness and tensile behavior. However, this is imperatively needed to exploit the potential of ADAM for customized design and implementation to develop advanced products. To address this issue, this research aims to investigate the fabrication parameters shell thickness and infill type, which have not been explored yet. In this respect, the effect of shell thickness and infill type on dimensional accuracy, mass/density, surface roughness and mechanical properties were explored. In addition to that, the microscopic structure of the specimens was investigated after tensile testing, and the obtained mechanical properties were compared to those of traditionally manufactured (wrought) specimens.

2. Materials and Methodology

2.1. Material

In 17-4 PH stainless steel (17-4PH SS), a well-known class of stainless steel, the hardening effect takes place due the martensitic precipitation. The 17–4PH SS powder is blended with wax and polymeric binder to give it a wire form and spooled in a coil. This was commercially acquired from the manufacturer (Desktop Metal Inc., Burlington, MA, USA). The metal powder composition (% wt.) was 0.07% C, 15.5–17.5% Cr, 3–5% Ni, 3–5% Cu, 1.0% Mn and 0.15–0.45% (Nb and Ta). The relatively high copper content provided the necessary precipitation towards the strengthening of this alloy [30].

2.2. Specimen Size

The specimen was a sub sized, 'dog-bone' shaped tensile testing specimen, following the code of ASTM E8/E8M [31]. 'Autodesk Inventor' software (Version 1.1) was used to draw the specimen in 3D format and export it as an STL file, which was readable by 'MarkForged' proprietary slicing software called 'Eiger'. 'Eiger' imported the CAD software and allow for calculations of all input parameters for the specimen to be fabricated. All settings were left at default, to permit the software to determine the optimum print settings, except shell thickness and infill pattern. The software provides a layer-by-layer breakdown of the printing process and the details of the infill for each part.

2.3. Input Parameters

In this research, the variable input parameters were infill type and shell thickness (top/bottom and walls). Two different infill types, namely, triangular and gyroid, were explored in this research. Figure 1 shows a comparison of gyroid and triangular infill types. The infill type dictates the structure of the internal pattern of the print inside the wall layers. Different infill structures require a different amount of material and time to create, therefore allowing for appropriate conclusions to be drawn regarding each infill's effect on the respective mechanical properties. Together with that, four different shell thicknesses were also explored, namely, 0.5, 1, 1.5 and 2 mm. Shell thickness refers to the thickness of the solid wall and top/bottom layers. Previous research conducted in-depth studies on layer thickness, but overall shell thickness was neglected; therefore, it was considered in this study. Figure 2 shows a comparison of shell thickness between two specimens.

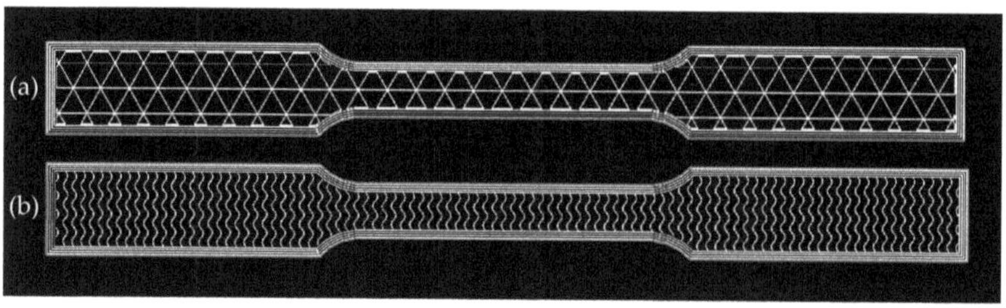

Figure 1. Triangular (**a**) and gyroid (**b**) infill pattern.

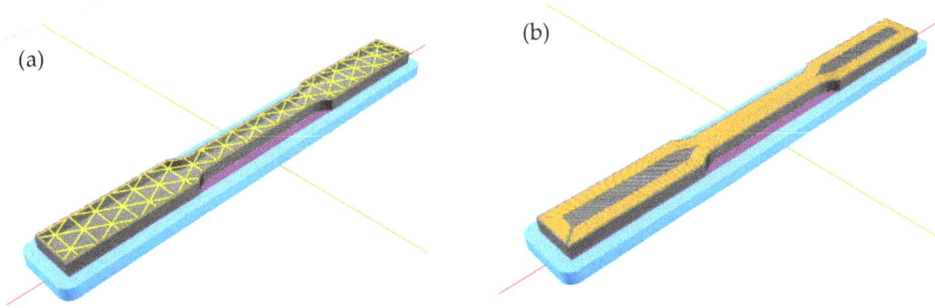

Figure 2. Visualization of shell thickness: 0.5 mm (**a**) and 3.5 mm (**b**).

Eight (8) unique 17-4 PH stainless-steel specimens were printed, each with varying attributes, as listed in Table 1. Each specimen was triplicated for tensile testing; thus, the total printed specimen number was 24.

Table 1. Specimen input parameter description.

Specimen Name	Shell Thickness (mm)	Infill Type
MF, GYR, 0.5 SHELL	0.5	Gyroid
MF, GYR, 1 SHELL	1	Gyroid
MF, GYR, 1.5 SHELL	1.5	Gyroid
MF, GYR, 2 SHELL	2	Gyroid
MF, TRI, 0.5 SHELL	0.5	Triangular
MF, TRI, 1 SHELL	1	Triangular
MF, TRI, 1.5 SHELL	1.5	Triangular
MF, TRI, 2 SHELL	2	Triangular

2.4. ADAM Process

The fabrication of the specimens incorporates three stages: 'printing', 'debinding' and 'sintering', before generating a final specimen. The filaments in wire form sat above the print bed. Before the start of a new print, the print bed was levelled; this ensured the initial layer of print was completed correctly, as it was a base for the rest of the parts. The Eiger slicing software connected with the printer, activating the printing process, which was then carried out over a time period defined by the input parameters for component composition and geometry. The printer was the commercially available MarkForged metal X^{TM}. Initially, a base plate known as a raft was printed onto the print bed, and then the ceramic binder was deposited before the initiation of the required specimen. The ceramic binder allows for sufficient strength during printing, debinding and sintering to hold the specimen in place on the print bed, while also allowing easy separation after sintering was complete to detach the specimen from the raft. Once printing was complete, the part was known as a 'green' part.

Printed specimens were placed into the stainless-steel basket before being lowered into the debinder unit. During this process, the binder that held together the metal particles dissolved, creating an 'open-pore' network. Once debinding was completed, the part was known as a 'brown' part.

The furnace was the location where 'atomic diffusion' occurs, fusing the metal particles together at elevated temperature to introduce strength and increase the mechanical properties of the printed specimen. The sintering time was around 24–30 h and constantly monitored. The maximum temperature achievable in the furnace was 1300 °C, which was below the melting point of printed materials (17-4 PH SS). Once the specimen was removed from the furnace, the rafts were easily removed by hand. The specimen, now called the 'white' part. could be used for its designed use or post-processing to further improve its physical and mechanical properties.

2.5. Measurement Instruments

The dimensional accuracy of the printed specimens was determined using a coordinate measuring machine (CMM) named Sheffield Discovery II D-8 CMM (Discovery II, manufactured by Sheffield, UK) accompanied by a diamond probe. This CMM was also used to measure the gauge thickness, gauge width, cylindricity at radii and warping of the specimen. The Mitutoyo SJ-201 surface roughness testing machine (Surftest Sj-210P, Mitutoyo, Tokyo, Japan) was used to measure surface roughness. A cut-off length of 25 mm was used. The hardness test used for this experiment was conducted using a Wilson 4JRa machine. Hardness results were recorded in HRB, three readings were taken per sample, and the data were averaged to increase reliability. The Shimadzu AGSX-300 (Shimadzu Autograph, Tokyo, Japan) was used to evaluate the tensile behavior of each specimen. The selected load cell was 300 kN, which was adequate for breaking steel and alloy specimens. A constant elongation rate of 1 mm/s was used on all specimens at ambient temperature (24 °C). The Olympus BX51M microscope (Olympus corporation, Tokyo, Japan) paired with Olympus stream motion software was used for optical microscopic evaluation of the specimens. The fractured tensile specimens were further examined by a field emission scanning electron microscope (FE-SEM), model Quanta 450 SEM (Thermofisher scientific, Waltham, MA, USA). The Archimedes principle was used to measure the density. A scale accurate to 2 decimal places, a beaker of demineralized water and a suspension stand with a piece of string for the specimen were used to measure the density of the specimen. The specimen was weighed in air, then subsequently attached to the string and weighed in water; the displacement of water was the weight taken as the buoyant force. Once the mass of displaced water was recorded, the density could be calculated.

3. Results and Discussion

3.1. Dimensional Accuracy

To access the dimensional accuracy of the printed specimens, four important measurements were taken, namely, cylindricity of gauge radius, height, perpendicularity and the overall length of the specimen. These were taken for every specimen printed after sintering.

3.1.1. Cylindricity Error at Gauge Radius

Figure 3 showed the average cylindricity error at gauge radius. The lower the value, the more cylindrical the radius was, with average value of 0.047 mm. The print heads were only able to move in an XY plane and not in a traditional smooth curve or arc form. The smaller resolution created more of an arc shape and less staircase effect [18].

The staircase effect, as shown schematically in Figure 4, was very prominent and depended on the line thickness in which the material was printed. This means that the thinner lines could conform to the arc boundaries of the printed part more accurately compared to those of the thicker lines. The line thickness depended on the nozzle diameter of the printer. To create an arc pattern, a very careful staircase effect had to be utilized. The employed printer, with a nozzle diameter of 50 μm, was not enough, as a perfect cylindrical formation was not achieved.

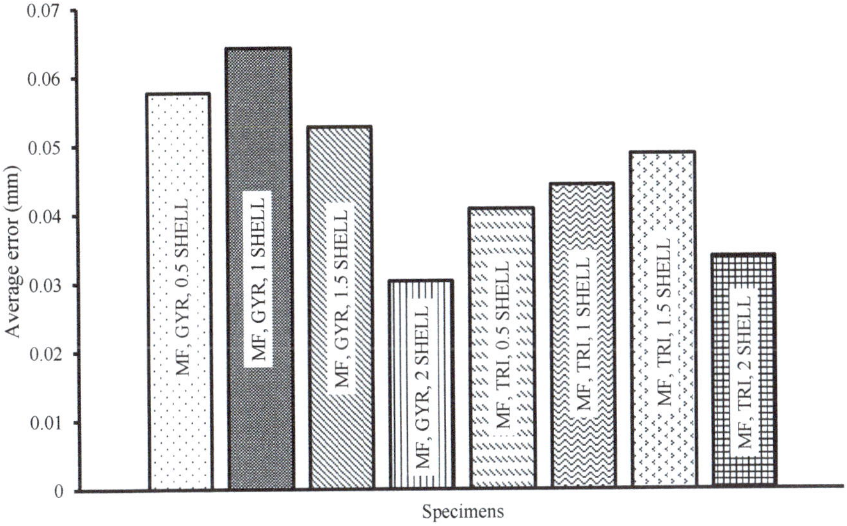

Figure 3. Cylindricity error in the specimens.

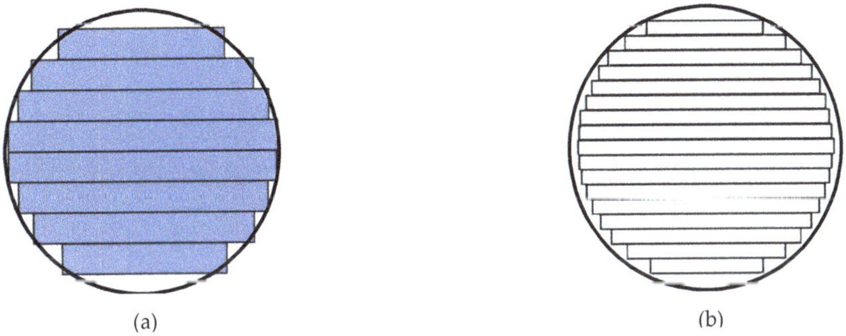

Figure 4. Staircase effect on 3D printed models: (**a**) thicker lines and (**b**) thinner lines.

3.1.2. Perpendicularity Error

Error in perpendicularity was shown in Figure 5. The error present in almost all specimens could be attributed to warping while sintering or print deformation during printing or sintering. During sintering, the metal was semi-melted and flowed to fill the holes/gaps in the microstructure. This could have resulted in a loss of dimensional accuracy, as they were free to move around.

3.1.3. Height Error

The vertical height error of each specimen is shown in Figure 6. The averaged error was 0.285 mm. Most specimens were close to the averaged values and showed great consistency in results. Once the specimen design was submitted to printer's proprietary software, the software enlarged the size by 20% to compensate shrinkage that occur during sintering. As the specimen was debinded, most of the binder material was dissolved away. In the furnace, the remaining binder was evaporated due to the heat, while the high temperature allowed for the sintering of metal powder, filling the porosity the binder had left behind. Galati et al. [6] also found a similar result when printing: different sized materials all exhibit a linear increase in size when printed, but after sintering, the larger specimens are known to be less close to the designed specimen size.

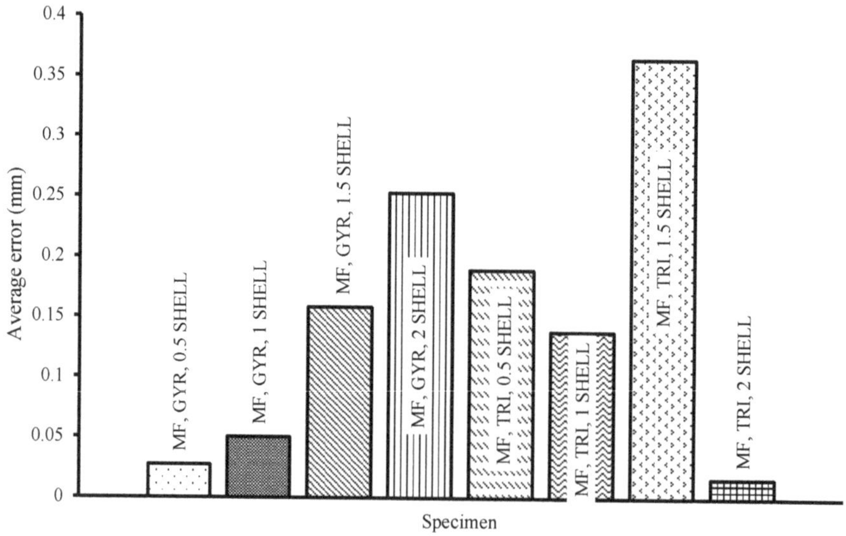

Figure 5. Perpendicularity error in the specimens.

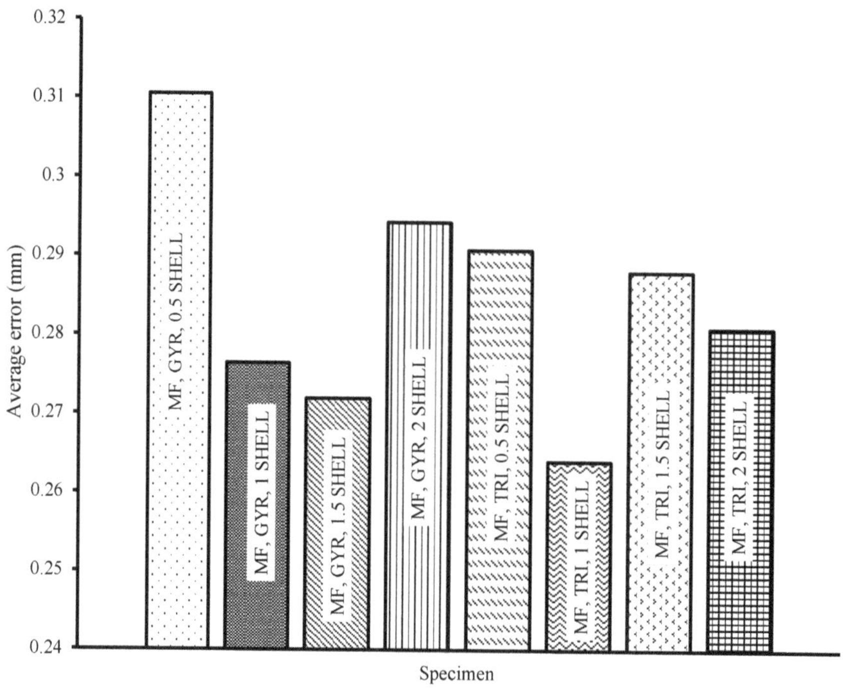

Figure 6. Vertical height error in the specimens.

3.1.4. Specimen Length Error

The overall length error of the specimen was shown in Figure 7, with an average value of 0.447 mm. All specimens were over the required dimensions, suggesting that there was a consistent issue that creates this discrepancy. A linear enlargement factor, as discussed earlier, could be a contributor to the oversized sintered specimens.

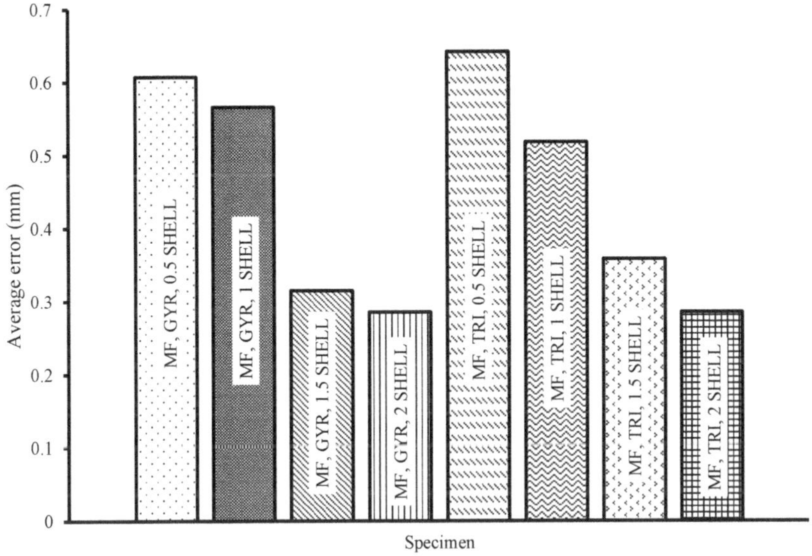

Figure 7. Length error in the specimens.

International tolerance (IT) grades were defined for ISO 286 [32], and the errors can be compared against IT standards. Table 2 below outlines IT grades for the standard ISO 286.

Table 2. ISO 286 IT grades [32].

Nominal Size (mm)		International Tolerance Grade (mm)					
>	≤	IT11	IT12	IT13	IT14	IT15	IT16
80	120	0.22	0.35	0.54	0.87	1.4	2.2

Thus, the printed specimens (100 mm) in the present case fall between IT13 and IT15. From the findings of Galati et al. [6], the dimensional accuracy and tolerance grades that were attained from ADAM were similar to the traditional process of forming and casting. The excess size error in these two processes allows for sufficient material to be removed using grinding or milling, just like traditional fabrication. Although a perfect end result was not achieved, it did limit the uses till further post-processing was performed to achieve the required tolerances for ADAM.

As can be seen from previously presented average error graphs (Figures 3 and 5–7), the gauge length error, perpendicularity error and height error of specimens stayed consistent throughout all print parameters, which may mean they had no correlation to infill type and shell thickness. The cylindricity error was lower for triangular infill specimens; however, the gauge length was higher for the triangular infill specimens. The overall length error of the specimen decreased with a lower shell thickness parameter. The maximum values for length error were recorded when the shell thickness was 0.5 mm, and the minimum values were when the shell thickness was 2 mm. When a lower shell thickness was used, less room for error was present, as the nozzle diameter stayed constant for every shell thickness that had to be printed. A larger shell thickness allowed for greater variation in printed thicknesses and also allowed for correction to occur if one wall was printed too thick.

3.2. Mass and Density of Printed Specimen

Table 3 shows the recorded mass of the specimens after sintering. The specimens exhibited a positive increase in mass with an increase in shell thickness for all specimens. Although the infill type showed minimal changes in mass among the specimens (less than 1.5% difference), the triangular infill was heavier most of the time. On average, increasing shell thickness from 0.5 to 1 mm caused a 32.20% mass increase; from 1 to 1.5 mm, a 20.16% mass increase; and from 1.5 to 2 mm, a 12.42% mass increase. The density of the printed specimens (after sintering) was 6745.47 kg/cm^3, which was close to the printer manufacturer's claimed value of 7440 kg/cm^3 [33].

Table 3. Masses of the printed specimens after sintering.

Specimen	Mass (g)
MF, GYR, 0.5 SHELL	19.21
MF, GYR, 1 SHELL	25.28
MF, GYR, 1.5 SHELL	30.19
MF, GYR, 2 SHELL	33.94
MF, TRI, 0.5 SHELL	19.07
MF, TRI, 1 SHELL	25.32
MF, TRI, 1.5 SHELL	30.62
MF, TRI, 2 SHELL	34.42

3.3. Surface Roughness

The measured surface roughness values for each specimen are shown in Figure 8. Although the arithmetic mean deviation (Ra) of the surface roughness is commonly used, the Rq value will be used for comparisons, as it is a better parameter to assess the average of surface roughness, as the graphs were sinusoidal in form. Thus, it was better to take an average that accounts for the root mean square (RMS), which Rq can. Rq may be used to determine whether the profile has projecting peaks that might have an effect on the function of the part [33]. The average Rq of the specimens was 5.27 μm on the top, whereas an average of 27.70 μm was recorded for the bottom. It was obvious that for all specimens, surface roughness at the bottom of the specimen was significantly higher when compared to the top surface. This was due to the presence of rafts at the bottom of the specimens, which were removed after printing. Traditional 3D printers printing polymeric components were able to achieve relatively smooth surface finishes [34] due to the absence of the sintering process. A smooth surface commonly had trends with good tensile strength, as crack propagation initially occurs at the surface of a specimen, where metal particles are not fully surrounded by other metal particles [35]. Miniature cracks can develop and cause larger cracks to propagate throughout the structure of the specimen, causing failure of the specimen at strengths much lower than predicted.

The specimens had an average side roughness (Rq) of 11.71 μm. Layer thickness had an important role in the surface roughness on the side of a specimen. Each layer was adjoined on top of the other, so if there were issues with the previous layer's adhesion, it would also impact the next one. Also, smaller layer thicknesses can have smaller valleys and peaks between each other in the horizontal orientation of the specimen, and layer height was dependent upon nozzle diameter.

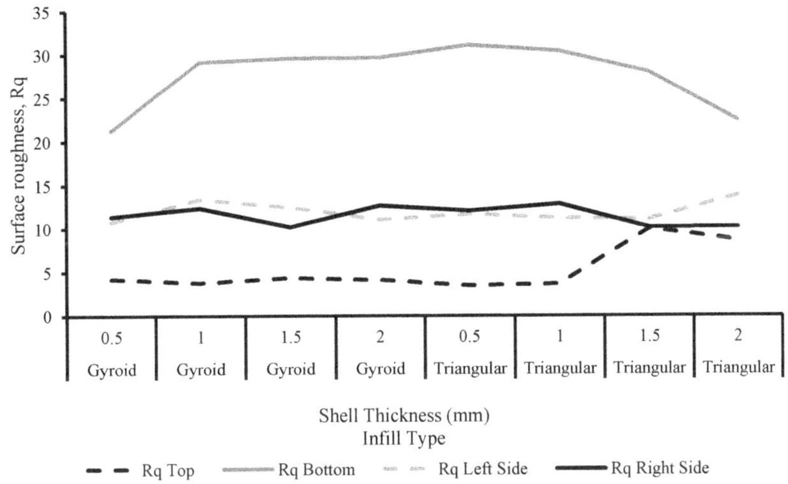

Figure 8. Effect of printing parameters on surface roughness (Rq).

3.4. Tensile Test

Stress–strain graphs of the specimens are plotted in Figure 9, together with a wrought [33] and manufacturer's claimed graph [33] for comparison purposes. The extracted mechanical properties of the specimens, like yield stress, ultimate tensile strength and strain % at break, are tabulated in Table 4. According to the manufacturer, the yield stress of the specimens was expected to achieve close to 800 MPa; however, results show a maximum recorded value of 252.9 MPa. The reason for this discrepancy was that the claimed values from the manufacturers were most likely conducted by a specimen with maximum solid-like structure, i.e., the maximum density of infill and shell thickness possible to create the strongest possible specimen. However, no fully solid specimens were created for this testing, as this was not required to assess shell thickness or infill type, and therefore, due to the presence of infill, the maximum potentiality of the mechanical properties might not have been able to be achieved. From Figure 9, it can be seen that there was a direct correlation between the increase in shell thickness and the increase in the specimen's ultimate tensile strength. Specimens that exhibit different infill types but have the same shell thickness tend to output a similar ultimate tensile strength. All specimens showed little or no signs of necking; brittle failure was quite frequently observed during tensile testing, with maximum elongation around 12.54% at break point. This result was expected, as 17-4 PH is known as a martensitic, precipitation-hardening stainless steel with brittle-like fracture appearance. Moreover, the ultimate tensile strength was about 4.88% lower.

It was not possible to directly compare the present data to those of the literature, as the printing parameters used to fabricate specimens in the present case were substantially different to those of the literature [17,21,23,26]. Having said that, data presented in the literature on the same material (17-4 PH stainless steel) system but different printing parameters were as follows: Burgess et al. [17] reported 395–1050 MPa of UTS and 28–37 HRC of hardness; Sambrook et al. [21] reported 550–730 MPa of UTS and 280–442 HV of hardness;

Thawon et al. [23] reported 809.89 MPa of UTS and 115.05–271.97 MPa of hardness; and Godec et al. [26] reported 9.95 MPa of UTS for the 'green parts' (before sintering) only.

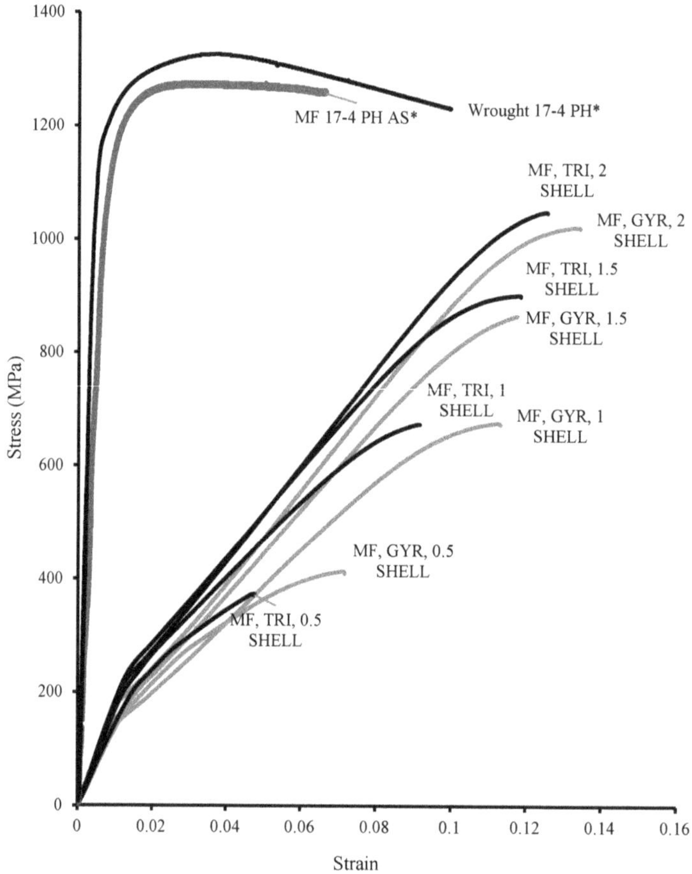

Figure 9. Stress–strain curves of the ADAM specimens with wrought and manufacturer's claimed curves * [33].

Table 4. Mechanical properties of the ADAM specimens together with wrought and manufacture's claimed values * [33].

Mechanical Properties	ADAM Specimen (MF, TRI, 2 SHELL)	Manufacture Datasheet *	Wrought Specimen *
Yield strength (MPa)	252.9	800	1000
Ultimate tensile strength (UTS, MPa)	1049.1	1050	1103
Strain at break (%)	12.54	5	5
Young's modulus (GPa)	15.6	140	200
Hardness (HRB)	290	277	322

The triangular infill pattern regularly outperformed the gyroid infill pattern in terms of ultimate tensile strength when the same shell thickness was compared (Figure 10). This could be due to the lack of support structure in the direction of load for the gyroid infill pattern. However, the gyroid infill pattern tended to have a longer elongation than the triangular infill. This may be due to the longer individual strands of the print present in the triangular infill pattern, whereas the gyroid infill had fewer length strands but a higher quantity of individual strands at the gauge area. The gyroid infill also had an alternating pattern of infill between every few printed layers, as seen in Figure 11. As the specimen was pulled apart by the tensile machine, the strands were assumed to be individually broken by having more strands present, and in alternative formation, a greater elongation distance was achieved by gyroid infill pattern.

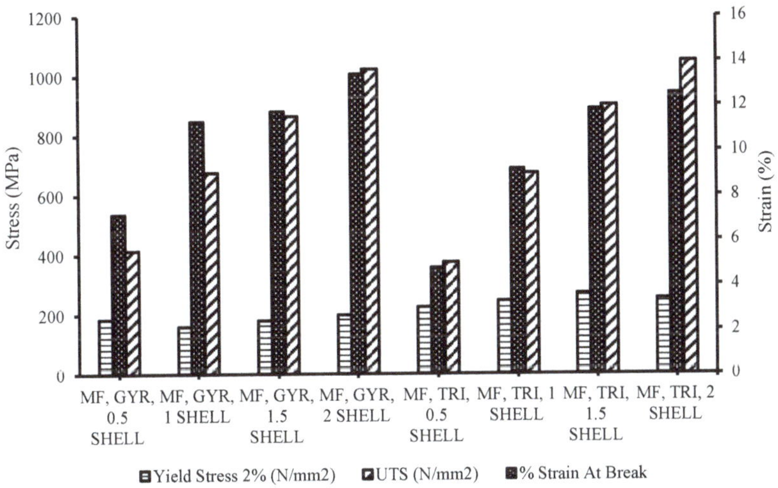

Figure 10. Comparison of the yield stress, ultimate tensile strength and strain % at break of the ADAM specimens.

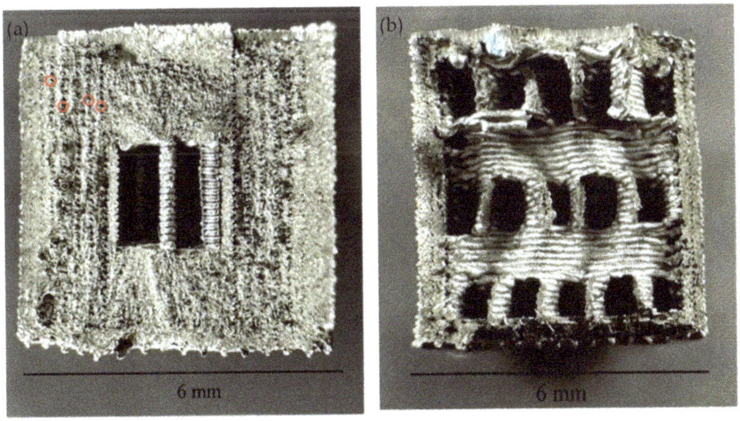

Figure 11. Morphology of (**a**) triangular and (**b**) gyroid infill pattern on the specimen after tensile testing, showing different fracture mode. The red circles indicate the presence of pores.

A strong relationship between measured mass and UTS was present, as shown in Figure 12. An increase in specimen mass directly translated to an increase in UTS. This was mainly due to the variation of shell thickness throughout the specimens. Shell thickness was a key component in specimen strength, and as there was more shell thickness available due to the extra mass of the specimen, the UTS increased accordingly. A 79.97% difference in sample mass was seen between the specimens, whereas a 180.44% difference was seen in UTS between the specimens. In summary, samples' internal geometries were influenced by the use of the specific infill type and shell thicknesses. Higher infill density and greater shell thickness obviously deposited more material and resulted in more mass compared to that of lower infill density and smaller shell thickness. This was corelated with the obtained mechanical properties (like YS and UTS), as reported in Figures 10 and 12.

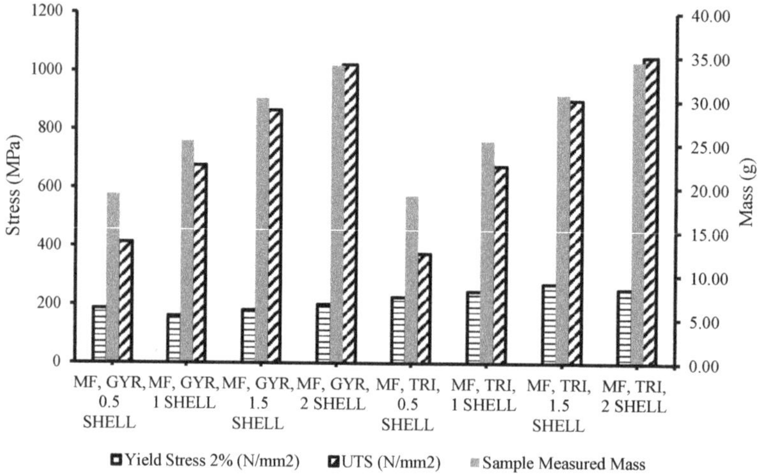

Figure 12. Yield stress, UTS and measured mass of the specimens.

3.5. Fracture Surface Morphology

The top-view and cross-section of the failed tensile test specimens were investigated further by optical and electron microscope to understand their fracture morphology. A representative optical microscope image (top-view) of a failed specimens is shown in Figure 13. Striation-like features were noticed and the shell wall, triangular infill pattern and fracture path were evident. The change of direction of the fracture path indicated that the crack was deflected in the interface of the shell and core of the specimen.

Representative cross-section fractography of the gyroid and triangular infill pattern specimens are shown in Figures 14 and 15, respectively. Irrespective of the specimens, it can be seen that there are two distinct zones, namely the shell wall and the inner core filled with either gyroid (Figure 14) or triangular (Figure 15) infill pattern. The shell wall represents complete fusion, withstanding voids formation among between the corners of the metal stands (Figures 14c and 15b,c). These pores/voids were present throughout the shell thickness, and their even spread conveys that this was a repetitive structure that formed during specimen printing. These were the typical void forms that occur due to incomplete fusion during the sintering process, as reported in the literature [7,23]. The core of the gyroid infill pattern specimen showed good fusion among the metal stands (Figure 14b), and the fracture mode was ductile fracture with representative dimple formation, together with 'cup and cone' appearance (Figure 14d). Strands of metal were laid down by the print nozzle and held together by a binder, which washed away during chemical debinding and sintering. However, the binder escaped through these holes, and sintering occurred where strand-to-strand diffusion took place in the metal particles. This was well known in the extrusion-based AM as a manufacturer's error that had not been addressed and led to porosity introduction in the samples [7,23].

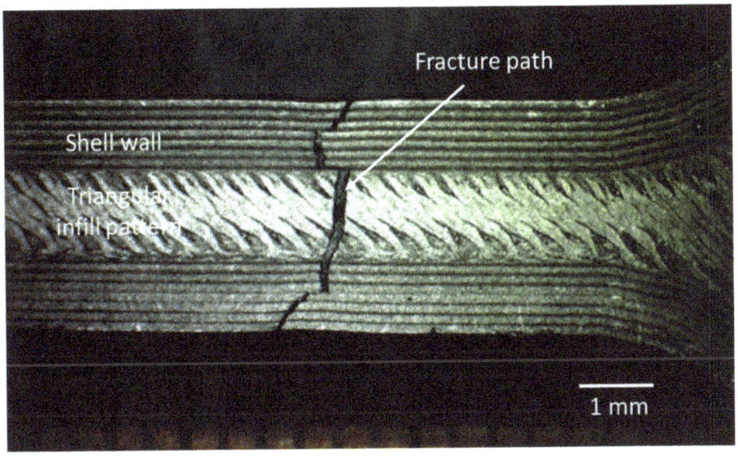

Figure 13. Striation-like features on the sample surface, together with fracture line and tool path (specimen: MF, TRI, 2 SHELL).

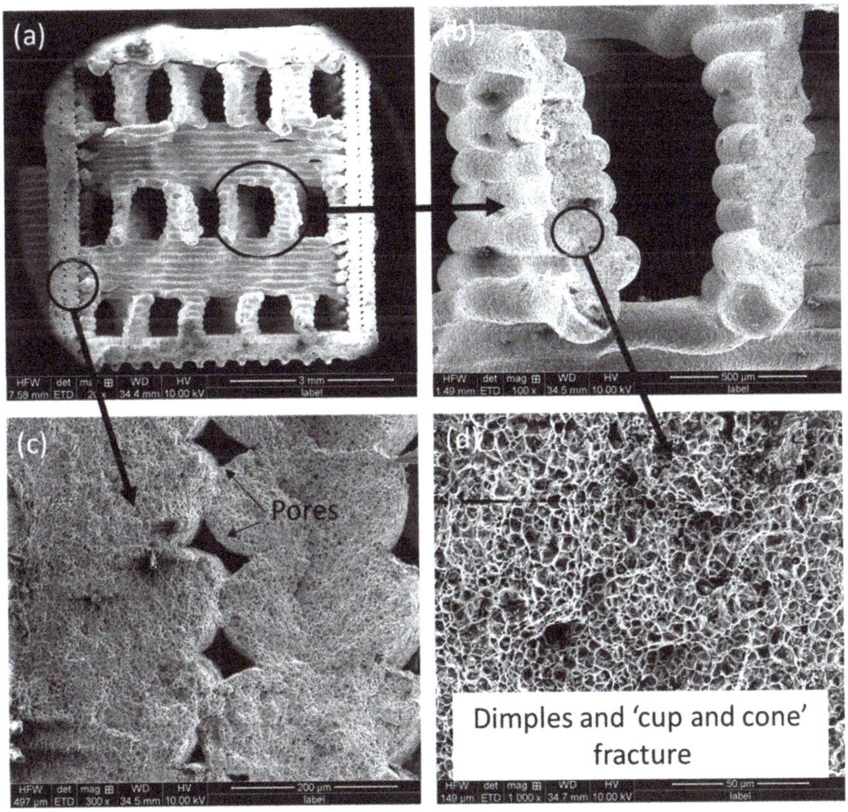

Figure 14. Fractography after tensile testing on a representative gyroid infill pattern specimen (MF, GYR, 0.5 SHELL): (**a**) overall fractured surface showing individual metal strands, (**b**) high-magnification view of metal strands, (**c**) pores among metal strands on shell wall and (**d**) dimples and 'cup and cone' representing ductile fracture.

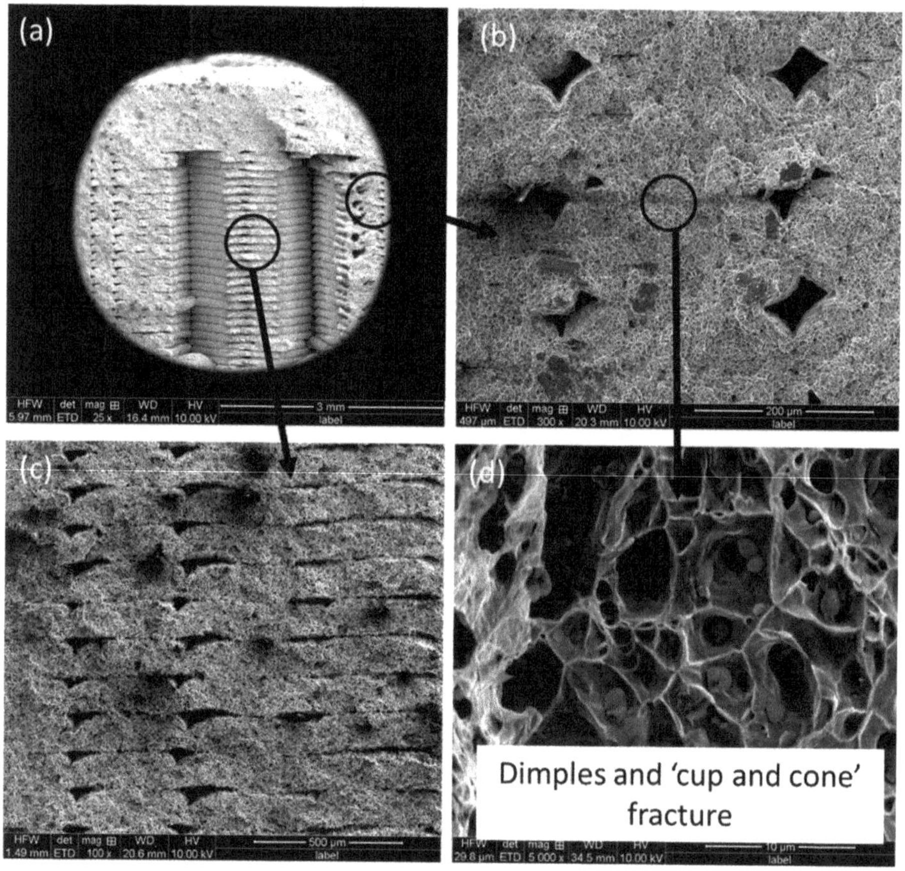

Figure 15. Fractography after tensile testing on representative triangular infill pattern specimen (MF, TRI, 2 SHELL): (**a**) overall fractured surface showing individual layers, (**b**) pores among shell wall layers, (**c**) pores among individual layers on the core of the specimen and (**d**) dimples and 'cup and cone' representing ductile fracture.

A higher magnified image (Figure 15d) on the triangular infill pattern specimen also demonstrates similar ductile fracture featuring dimples and 'cup and cone', together with unmelted particles.

The feed stock during the printing process was metal powder with polymer binder. The polymer binder burnt out during the sintering process, which gave rise to various gasses such as carbon dioxide and hydrogen and shoots like carbon due to the lack of oxygen in the argon-rich environment [36]. These trapped gases not only gave rise to pores but also voids, particularly in the corners of the metal strands, due to "inadequate sintering time or temperature, and quite possibly insufficient chemical de-binding time" [37]. The shell wall of the specimens first heated up during the sintering process and fused the metal layers. As the process continued, the heat continued to flow towards the inner core of the specimen where it became fused and densified. The evidence in Figures 14 and 15 suggests that the inner core of the specimens was not fused completely, which might be related to lack of sintering time, temperature or both. The whole sintering process and associated parameters were proprietary property of the manufacture, and thus, they were beyond the scope of adjustment, which required further refinement.

4. Conclusions

The present investigation reports on the effect of varying input parameters on the dimensional accuracy and tensile properties on specimens manufactured by ADAM. From the observed results and analysis, the following conclusions could be made:

1. Results from the coordinate measuring machine showed there were inconsistencies in the print quality of the ADAM processes. Toolpath geometry, staircasing and sintering technology could be reasons for the discrepancies; however, when compared with traditional casting and forging of similar steel, international tolerance grades can be achieved. Overall length error was at a maximum when shell thickness was at a minimum value. Triangular infill specimens saw a decrease in cylindricity error and an increase in gauge length error when compared to gyroid fill. All of these errors were in an acceptable range by the international tolerance (IT) grades of ISO 286.
2. A minimal difference in mass (1.5%) was observed when the infill type was changed. A 21.6% increase in specimens' mass was observed when shell thickness was increased in each iteration. Parts printed by ADAM can have up to ¼ mass savings when compared to traditional manufacturing, at the expense of the parts' strength.
3. The specimens showed comparable ultimate tensile strength (1049.1 MPa), matching the claims of the manufacturer (1050 MPa), together with elongation at break, though they were about 4.88% lower in ultimate tensile strength. The specimens showed an increase in ultimate tensile strength when the shell thicknesses of specimens were increased. The two different infill types saw minimal changes, although it should be noted that triangular specimens exhibited greater ultimate tensile strength, whereas the gyroid had slightly longer elongation at break.
4. Microscopic analysis of specimens showed deformities that include striations from tool path error. The cross-section of the tensile tested, broken specimen revealed significant pores in the microstructure and could contribute to a reduction in the mechanical properties of the specimens. Thus, further optimization of the sintering time and temperature was foreseen.

ADAM is a growing and developing technique in its early stages. Outcomes from its processes now may be extremely different from their outputs in the near future with greater access to technology and research. Many input parameters are able to be tested, but this investigation only considered two. Countless more can be varied to ultimately refine and optimize print parameters for maximum quality of print features and properties for ADAM.

Author Contributions: A.K.B.: conceptualization, experiment, writing—review and editing; J.M.S.: experiment, formal analysis, validation, data curation, writing—review and editing; A.P.: resource, conceptualization, supervision, writing—review and editing. All authors have read and agreed to the published version of the manuscript.

Funding: This research received no external funding.

Data Availability Statement: The raw/processed data used to produce the results will be made available by the corresponding author upon reasonable request.

Conflicts of Interest: The authors declare that they have no known competing financial interests or personal relationships whatsoever that could have influenced the work reported in this paper.

References

1. Henry, T.C.; Morales, M.A.; Cole, D.P.; Shumeyko, C.M.; Riddick, J.C. Mechanical behavior of 17-4 PH stainless steel processed by atomic diffusion additive manufacturing. *Int. J. Adv. Manuf. Technol.* **2021**, *114*, 2103–2114. [CrossRef]
2. Basak, A.K.; Celis, J.-P.; Vardavoulias, M.; Matteazzi, P. Effect of nanostructuring and Al alloying on friction and wear behaviour of thermal sprayed WC–Co coatings. *Surf. Coat. Technol.* **2012**, *206*, 3508–3516. [CrossRef]
3. Basak, A.; Matteazzi, P.; Vardavoulias, M.; Celis, J.-P. Corrosion-wear behaviour of thermal sprayed nanostructured FeCu/WC–Co coatings. *Wear* **2006**, *261*, 1042–1050. [CrossRef]
4. Bajaj, P.; Hariharan, A.; Kini, A.; Kürnsteiner, P.; Raabe, D.; Jägle, E.A. Steels in additive manufacturing: A review of their microstructure and properties. *Mater. Sci. Eng. A* **2020**, *772*, 138633. [CrossRef]
5. Attaran, M. The rise of 3-D printing: The advantages of additive manufacturing over traditional manufacturing. *Bus. Horiz.* **2017**, *60*, 677–688. [CrossRef]
6. Galati, M.; Minetola, P. Analysis of density, roughness, and accuracy of the atomic diffusion additive manufacturing (ADAM) process for metal parts. *Materials* **2019**, *12*, 4122. [CrossRef]
7. Basak, A.; Lee, A.; Pramanik, A.; Neubauer, K.; Prakash, C.; Shankar, S. Material extrusion additive manufacturing of 17-4 PH stainless steel: Effect of process parameters on mechanical properties. *Rapid Prototyp. J.* **2023**, *29*, 1097–1106. [CrossRef]
8. Bjørheim, F.; Lopez, I.L.T. Tension testing of additively manufactured specimens of 17-4 PH processed by Bound Metal Deposition. In *IOP Conference Series: Materials Science and Engineering*; IOP Publishing: Bristol, UK, 2021; p. 12037.
9. Gonzalez-Gutierrez, J.; Cano, S.; Schuschnigg, S.; Kukla, C.; Sapkota, J.; Holzer, C. Additive manufacturing of metallic and ceramic components by the material extrusion of highly-filled polymers: A review and future perspectives. *Materials* **2018**, *11*, 840. [CrossRef]
10. Gabilondo, M.; Cearsolo, X.; Arrue, M.; Castro, F. Influence of build orientation, chamber temperature and infill pattern on mechanical properties of 316l parts manufactured by bound metal deposition. *Materials* **2022**, *15*, 1183. [CrossRef]
11. Degnah, A.; Tabbakh, T.; Kurdi, A.; Basak, A.K. Role of precipitation and solute segregation on micro-scale deformation of additively manufactured Inconel 718. *Mater. Sci. Eng. A* **2023**, *887*, 145762. [CrossRef]
12. Becker, T.H.; Kumar, P.; Ramamurty, U. Fracture and fatigue in additively manufactured metals. *Acta Mater.* **2021**, *219*, 117240. [CrossRef]
13. Tabbakh, T.; Alshihri, S.; Basak, A.; Kurdi, A. Strength of a 3D printed Al 7068 alloy under micro-pillar compression. *Met. Mater. Int.* **2022**, *28*, 2706–2718. [CrossRef]
14. Ngo, T.D.; Kashani, A.; Imbalzano, G.; Nguyen, K.T.; Hui, D. Additive manufacturing (3D printing): A review of materials, methods, applications and challenges. *Compos. Part B Eng.* **2018**, *143*, 172–196. [CrossRef]
15. Kurdi, A.; Basak, A. Micro-mechanical behaviour of selective laser melted Ti6Al4V under compression. *Mater. Sci. Eng. A* **2021**, *826*, 141975. [CrossRef]
16. Chacón, J.M.; Caminero, M.A.; García-Plaza, E.; Núnez, P.J. Additive manufacturing of PLA structures using fused deposition modelling: Effect of process parameters on mechanical properties and their optimal selection. *Mater. Des.* **2017**, *124*, 143–157. [CrossRef]
17. Burgess, A.; Dodd, R.; Radwani, M.; Opoz, T.; Tammas-Williams, S. The properties of stainless steel 17-4PH produced by a low-cost additive manufacturing technique, the Markforged MetalX™. *J. Phys. Conf. Ser.* **2022**, *2198*, 12057. [CrossRef]
18. Vispute, M.; Kumar, N.; Taufik, M.; Jain, P.K. Improving surface finish of extrusion based additive manufactured parts using novel triangle based toolpath approach. *Int. J. Interact. Des. Manuf.* **2024**, *18*, 433–452. [CrossRef]
19. Ajay, V.; Nakrani, J.; Mishra, N.K.; Shrivastava, A. Anisotropic fatigue crack propagation in wire arc additively manufactured 316L stainless steel. *Int. J. Fatigue* **2023**, *177*, 107976. [CrossRef]
20. Sambrook, C. The Effect of Heat Treatment and Print Direction on Additively Manufactured 17-4 PH Stainless Steel. Master's Thesis, The University of Greensland, St. Lucia, Australia, 2021.
21. Rosnitschek, T.; Seefeldt, A.; Alber-Laukant, B.; Neumeyer, T.; Altstädt, V.; Tremmel, S. Correlations of geometry and infill degree of extrusion additively manufactured 316l stainless steel components. *Materials* **2021**, *14*, 5173. [CrossRef]
22. Thawon, I.; Fongsamootr, T.; Mona, Y.; Suttakul, P. Investigation of the mechanical properties of additively manufactured metal parts with different relative densities. *Appl. Sci.* **2022**, *12*, 9915. [CrossRef]
23. Basak, A.K.; Pramanik, A.; Chen, Y.X.; Prakash, C.; Radhika, N.; Shankar, S. Bound metal deposition of stainless steel 316L: Effect of process variables on microstructural and mechanical behaviors. *Materialia* **2024**, *36*, 102196. [CrossRef]
24. Godec, D.; Cano, S.; Holzer, C.; Gonzalez-Gutierrez, J. Optimization of the 3D printing parameters for tensile properties of specimens produced by fused filament fabrication of 17-4PH stainless steel. *Materials* **2020**, *13*, 774. [CrossRef] [PubMed]

25. Shakeri, Z.; Benfriha, K.; Shirinbayan, M.; Ahmadifar, M.; Tcharkhtchi, A. Mathematical modeling and optimization of fused filament fabrication (FFF) process parameters for shape deviation control of polyamide 6 using Taguchi method. *Polymers* **2021**, *13*, 3697. [CrossRef] [PubMed]
26. Raju, N.; Warren, P.; Subramanian, R.; Ghosh, R.; Raghavan, S.; Fernandez, E.; Kapat, J. Material Properties of 17-4PH Stainless Steel Fabricated by Atomic Diffusion Additive Manufacturing (ADAM). In Proceedings of the 32nd Annual International Solid Freeform Fabrication Symposium—An Additive Manufacturing Conference, Austin, TX, USA, 2–4 August 2021.
27. Rodriguez, J.; Zuriarrain, A.; Madariaga, A.; Arrazola, P.J.; Dominguez, E.; Fraile, I.; Soler, D. Mechanical Properties and Fatigue Performance of 17-4 PH Stainless Steel Manufactured by Atomic Diffusion Additive Manufacturing Technology. *J. Manuf. Mater. Process.* **2023**, *7*, 172. [CrossRef]
28. Bouaziz, M.A.; Djouda, J.M.; Kauffmann, J.; Hild, F. Microscale mechanical characterization of 17-4PH stainless steel fabricated by Atomic Diffusion Additive Manufacturing (ADAM). *Procedia Struct. Integr.* **2020**, *28*, 1039–1046. [CrossRef]
29. Bouaziz, M.; Djouda, J.M.; Chemkhi, M.; Rambaudon, M.; Kauffmann, J.; Hild, F. Heat treatment effect on 17-4PH stainless steel manufactured by Atomic Diffusion Additive Manufacturing (ADAM). *Procedia CIRP* **2021**, *104*, 935–938. [CrossRef]
30. Hsiao, C.; Chiou, C.; Yang, J. Aging reactions in a 17-4 PH stainless steel. *Mater. Chem. Phys.* **2002**, *74*, 134–142. [CrossRef]
31. *ASTM E8/E8M-16a*; Standard Test Methods for Tension Testing of Metallic Materials. ASTM International: West Conshohocken, PA, USA, 2016.
32. *ISO 286-1: 2010*; Geometrical Product Specifications (GPS)—ISO Code System for Tolerances on Linear Sizes. International Organization for Standardization: Geneva, Switzerland, 2010.
33. Markforged. Available online: https://markforged.com/materials/metals/17-4-ph-stainless-steel (accessed on 21 May 2025).
34. Nigam, A.; Tai, B.L. Effects of in-process surface finishing on part strength in polymer material extrusion additive manufacturing. *Addit. Manuf.* **2024**, *80*, 103960. [CrossRef]
35. Purnowidodo, A.; Arief, S.S.; Iman, F.H. Effect surface roughness on fatigue crack propagation behaviour of fibre Metal Laminates (FMLs). *Int. J. Automot. Mech. Eng.* **2019**, *16*, 6588–6604. [CrossRef]
36. Burge, S.J.; Tipper, C. The burning of polymers. *Combust. Flame* **1969**, *13*, 495–505. [CrossRef]
37. Thompson, Y.; Zissel, K.; Förner, A.; Gonzalez-Gutierrez, J.; Kukla, C.; Neumeier, S.; Felfer, P. Metal fused filament fabrication of the nickel-base superalloy IN 718. *J. Mater. Sci.* **2022**, *57*, 9541–9555. [CrossRef] [PubMed]

designs

MDPI

Article

Crash Performance of Additively Manufactured Tapered Tube Crash Boxes: Influence of Material and Geometric Parameters

Ahmed Saber [1], Mehmet Ali Güler [1,*], Erdem Acar [2], Omar Soliman ElSayed [1], Hussain Aldallal [1], Abdulrahman Alsadi [1] and Yousef Aldousari [1]

[1] College of Engineering and Technology, American University of the Middle East, Egaila 54200, Kuwait; ahmed.saber@aum.edu.kw (A.S.); 66421@aum.edu.kw (O.S.E.); 66246@aum.edu.kw (H.A.); 57186@aum.edu.kw (A.A.); 59751@aum.edu.kw (Y.A.)

[2] Department of Mechanical Engineering, TOBB University of Economics and Technology, Ankara 06560, Türkiye; erdem.acar@gmail.com

* Correspondence: mehmet.guler@aum.edu.kw

Abstract: Crash boxes play a crucial role in mitigating force during vehicle collisions by absorbing impact energy. Additive manufacturing (AM), particularly Fused Deposition Modeling (FDM), has emerged as a promising method for their fabrication due to its design flexibility and continuous advancements in material development. This study investigates the crash performance of tapered crash box configurations, each manufactured using two FDM materials: Carbon Fiber-Reinforced Polylactic Acid (PLA-CF) and Polylactic Acid Plus (PLA+). The specimens vary in wall thickness and taper angles to evaluate the influence of geometric and material parameters on crashworthiness. The results demonstrated that both specific energy absorption (SEA) and crush force efficiency (CFE) increase with wall thickness and taper angle, with PLA-CF consistently outperforming PLA+ in both metrics. ANOVA results showed that wall thickness is the most influential factor in crashworthiness, accounting for 73.18% of SEA variation and 58.19% of CFE variation. Taper angle contributed 13.49% to SEA and 31.49% to CFE, while material type had smaller but significant effects, contributing 0.66% to SEA and 0.11% to CFE. Regression models were developed based on the experimental data to predict SEA and CFE, with a maximum absolute percentage error of 4.97%. These models guided the design of new configurations, with the optimal case achieving an SEA of 32.086 ± 0.190 kJ/kg and a CFE of 0.745 ± 0.034. The findings confirm the potential of PLA-CF in enhancing the energy-absorption capability of crash boxes, particularly in tapered designs.

Keywords: additive manufacturing; 3D printing; crashworthiness; energy absorption; tapered tubes; crash box

1. Introduction

Crash boxes are critical components designed to absorb impact energy during vehicle collisions, reducing the force transmitted to the main structure and enhancing occupant safety. They are specifically engineered to deform upon impact, effectively dissipating energy. Crashworthiness refers to a crash box's ability to absorb impact energy and protect occupants during a collision [1]. Traditionally, crash boxes have been manufactured using processes such as composite fabrication [2], wire electrical discharge machining [3], extrusion [4,5], welding [6,7], CNC water jet cutting [8], and stamping [9,10]. Crash boxes fabricated using these methods often involve complex, multi-step production processes and offer limited geometric flexibility. In recent years, additive manufacturing (AM), originally introduced by Charles Hull in 1984 [11], has significantly expanded design possibilities by enabling the fabrication of intricate geometries previously unattainable with conventional techniques. This advancement is particularly important for crash box development, where geometry plays a central role in energy-absorption performance.

A variety of additive manufacturing (AM) techniques are available, each offering distinct advantages and limitations. The selection of an appropriate AM method depends on several factors, including material compatibility, processing speed, resolution, and cost-effectiveness. AM technologies such as Stereolithography (SLA) [12,13], Material Jetting (MJT) [14,15], Selective Laser Melting (SLM) [16,17], Selective Laser Sintering (SLS) [18,19], and Fused Deposition Modeling (FDM) [20–22] have been successfully employed in the fabrication of crash boxes, demonstrating their capability to produce complex geometries with tailored energy-absorption characteristics. Among various AM techniques, FDM has gained widespread adoption due to its accessibility, versatility, and advancements in both technology and material development, making it increasingly efficient, cost-effective, and suitable for crash box production [23]. Additionally, FDM enables the fabrication of fiber-reinforced polymer structures that offer an optimal balance of lightweight properties, structural integrity, and design flexibility, making them well-suited for lightweight energy-absorbing applications.

Several studies have compared different materials used in FDM to assess their impact on energy-absorbing performance [24–26]. For example, Zhang et al. [27] investigated metamaterial lattice structures fabricated with Polylactic Acid (PLA) and Polyethylene Terephthalate Glycol (PETG). Their study examined unit-cell configurations, including hexagonal, hybrid, and re-entrant layouts, with PETG re-entrant honeycomb demonstrating the highest performance. Similarly, Isaac et al. [28] investigated the performance of five polymer-based honeycomb lattice structures fabricated from four different materials: PLA, PETG, ABS (Acrylonitrile Butadiene Styrene), ASA (Acrylonitrile Styrene Acrylate), and PA-CF (Polyamide–Carbon Fiber), highlighting PETG's superior energy-absorption performance. Furthermore, Wang et al. [29] explored the energy-absorption properties of thin-walled tubular structures made from PA (Polyamide), PA-CF, and PA-GF (Polyamide–Glass Fiber). They concluded that PA-CF exhibited the highest energy-absorption performance.

In addition to material considerations, the geometric configurations enabled by FDM technology have been extensively investigated to enhance energy-absorption performance [30–32]. Wang et al. [33] examined multi-cell-filled tubes with internal cell geometries, including circular, hexagonal, and triangular shapes, all fabricated from PA-CF. They concluded that circular-filled configurations demonstrated superior energy absorption to both hexagonal and triangular designs. Liu et al. [34] explored stepwise graded multi-cell tubes (SGMTs) and continuous graded multi-cell tubes (CGMTs), produced using PA-CF. Their findings showed that CGMTs offered improved energy absorption due to the absence of discontinuous interfaces, which contributed to more uniform deformation. Liu et al. [35] investigated assembled and integrated lattice-filled multi-cell tubes made from PA-CF. Their integrated designs exhibited enhanced performance, benefiting from a synergistic effect between the lattice and tube structures, resulting in greater energy absorption than the sum of their individual components.

The continuous advancement of Fused Deposition Modeling (FDM) materials, particularly Carbon Fiber-Reinforced Polylactic Acid (PLA-CF) and Polylactic Acid Plus (PLA+), has opened new avenues for fabricating crash boxes with enhanced energy-absorption capabilities. These materials were selected based on their commercial availability, compatibility with standard FDM printers, and suitability for lightweight structural applications without requiring specialized processing. PLA-CF offers improved stiffness and strength due to its carbon fiber reinforcement, while PLA+ delivers enhanced ductility and toughness through its modified formulation. Although PLA-CF demonstrates higher stiffness, crashworthiness is influenced by a combination of material and geometric characteristics, including ductility, deformation behavior, and energy dissipation mechanisms. Therefore, both materials present distinct advantages for energy-absorbing structures. However, a systematic comparison between these two materials has not yet been explored. In this study, the crash performance of nine distinct tapered-tube crash box designs was investigated, with each design printed once using PLA+ and once using PLA-CF, resulting in a total of 18 specimens. The specimens featured varying wall thicknesses and tapered angles to comprehensively assess their effects on crash performance. Upon completing the testing of 18 specimens, the best candidates were further investigated by varying the thickness and the taper angle. The additional three experiments concluded the investigation. This study highlights how material and geometric parameters together affect crashworthiness, aiding the design of more effective FDM-manufactured crash boxes. The results of this study can serve as benchmark results for finite element analysis comparison.

2. Experimental Procedure

2.1. Geometrical Design

A tapered-tube geometry was adopted in this study to evaluate and compare the crashworthiness performance of different FDM materials, as tapered configurations have demonstrated superior energy-absorption characteristics to straight tubes [36]. All the specimens maintained a constant height (H) of 60 mm and an outer base diameter (D) of 30 mm, where T denotes the wall thickness and A represents the taper angle. To systematically investigate the influence of geometrical parameters, three wall thicknesses (1.0 mm, 1.5 mm, and 2.0 mm) and three taper angles (0°, 2.5°, and 5.0°) were selected. This full factorial design resulted in nine distinct geometric configurations. Each configuration was fabricated once using PLA+ and once using PLA-CF, yielding a total of 18 specimens for crash testing. A 50 mm × 50 mm × 2 mm base plate was integrated into the bottom of each specimen to ensure stability during compression testing. The geometry and dimensional specifications of the specimens are illustrated in Figure 1, while the specific configurations of wall thickness and taper angle used in the study are summarized in Table 1.

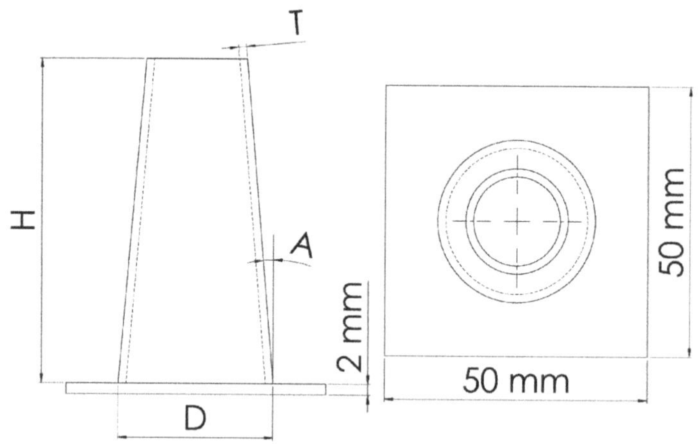

Figure 1. Geometry and dimensional specifications of the specimens.

Table 1. Design configurations based on T and A.

Configuration	T [mm]	A [°]
Config 1	1	0
Config 2	1	2.5
Config 3	1	5
Config 4	1.5	0
Config 5	1.5	2.5
Config 6	1.5	5
Config 7	2	0
Config 8	2	2.5
Config 9	2	5

2.2. Materials and Processing Parameters

In this study, all the specimens were fabricated using a 1.75 mm filament provided by Shenzhen eSUN Industrial Co., Ltd., Shenzhen, China, with the corresponding mechanical properties summarized in Table 2. The geometrical models were created using SOLIDWORKS and exported in STL (Standard Tessellation Language) format. These STL files were processed in PrusaSlicer to generate G-code instructions for the Prusa XL FDM 3D printer, both developed and manufactured by Prusa Research, based in Prague, Czech Republic. FDM-fabricated components exhibit anisotropic mechanical properties due to differences in inter-layer adhesion. To ensure consistency and isolate the influence of geometry and material, all specimens were printed in a vertical orientation (Z-axis), with layers parallel to the base plate. This orientation was selected, as it offers enhanced structural performance under axial loading and is commonly adopted in studies evaluating crash performance [37]. The printing parameters were selected following the recommendations of both the filament and printer manufacturers. The applied printing parameters are detailed in Table 3.

Table 2. Mechanical properties of PLA+ and PLA-CF.

Property	PLA+	PLA-CF
Density (g/cm^3)	1.23	1.21
Tensile Strength (MPa)	63	39
Elongation at Break (%)	20	4.27
Flexural Strength (MPa)	74	103
Flexural Modulus (MPa)	1973	5003
IZOD Impact Strength (kJ/m^2)	9	5.08

Table 3. Printing parameters of PLA+ and PLA-CF.

Parameter	PLA+	PLA-CF
Printing temperature (°C)	220	220
Bed temperature (°C)	60	60
Diameter of nozzle (mm)	0.4	0.4
Layer height (mm)	0.15	0.15
Infill density (%)	100	100
Outer wall print speed (mm/s)	80	80
Inner wall print speed (mm/s)	110	110

Figure 2a displays the final printed specimens, each labeled on its integrated base for identification. These alphanumeric labels denote specific geometrical and material attributes; for instance, "T2A0C" corresponds to a 2 mm wall thickness, 0° taper angle, and PLA-CF material, while "T1A2.5+" represents a 1 mm wall thickness, 2.5° taper angle, and PLA+ material. For each PLA+ and PLA-CF specimen, a separate base (50 mm × 50 mm × 2 mm) was 3D printed, as illustrated in Figure 2b.

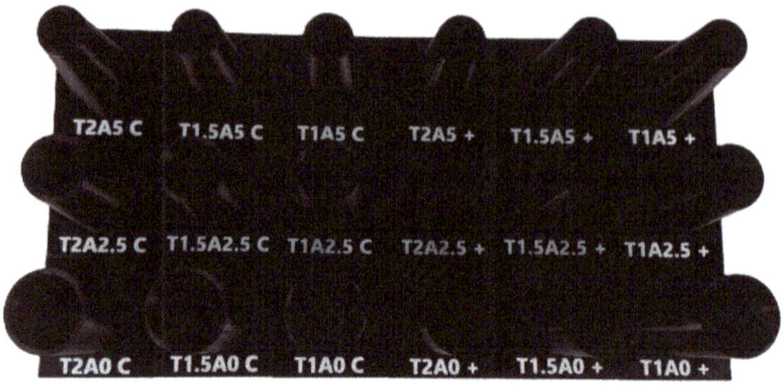

(**a**)

(**b**)

Figure 2. Three-dimensional printed (**a**) PLA+ and PLA-CF specimens; (**b**) PLA+ and PLA-CF bases.

2.3. Experimental Method

Quasi-static axial compression tests were conducted at room temperature to assess the crashworthiness of the fabricated specimens. A Tinius Olsen universal testing machine (Model 50ST), manufactured by Tinius Olsen (Horsham, PA, USA) and equipped with a 50 kN load cell, was used for testing, as shown in Figure 3. Prior to testing, the specimen's total mass, including the base, was measured, and the mass of the base was subtracted to determine the mass of the specimen alone. The specimens were placed between the two flat plates of the testing machine. The lower plate incorporated a 3D-printed frame with a groove (50 mm × 50 mm × 2 mm) that securely held the specimen base in position, ensuring proper alignment and preventing slippage during compression, as shown in Figure 3. The upper plate moved downward at a speed of 5 mm/min [38,39], applying a compressive force on the specimens, with the crushing displacement set to 2/3 of the specimen's original height [33,40]. Force-displacement data were continuously recorded during the compression tests.

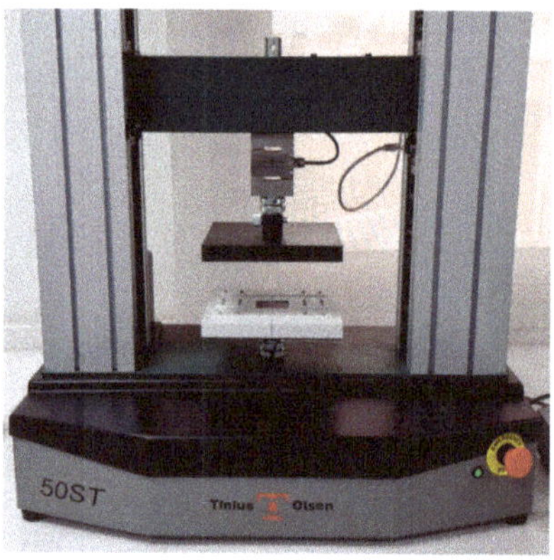

Figure 3. Tinius Olsen universal testing machine used in the testing of specimens.

2.4. Definition of Crashworthiness Indicators

Crash box performance is typically evaluated through key metrics that indicate its ability to absorb impact forces. These metrics are derived from the force-displacement curve. The primary parameters used for this evaluation are summarized below [41]:

2.4.1. Total Energy Absorption, E_T

The total energy absorption (E_T) of a crash box can be determined by calculating the work achieved by the crushing force. It is represented by the area under the axial force versus the axial displacement curve. E_T is expressed as follows:

$$E_T = \int_0^{\delta_{max}} F d\delta \quad (1)$$

where F is the crushing force, δ is the displacement, and δ_{max} is the total crush displacement.

2.4.2. Peak Crush Force, PCF

The peak crush force (PCF) is the maximum force observed in the axial direction during the crushing process.

2.4.3. Mean Crush Force, MCF

The mean crush force (MCF) is defined as the total energy absorbed per unit of total crush displacement. It is expressed as follows:

$$MCF = \frac{\int_0^{\delta_{max}} F d\delta}{\delta_{max}} \quad (2)$$

2.4.4. Specific Energy Absorption, SEA

Specific energy absorption (SEA) is defined as the total energy absorbed per unit mass of the crash box. It is expressed as follows:

$$SEA = \frac{E_T}{m} \quad (3)$$

where m is the mass of the crash box. A higher SEA indicates greater energy absorption per unit mass, which helps reduce the kinetic energy transmitted to occupants and thereby enhances safety.

2.4.5. Crush Force Efficiency, CFE

Crush force efficiency (CFE) is defined as the ratio of the MCF to the PCF. It is expressed as follows:

$$CFE = \frac{MCF}{PCF} \quad (4)$$

A higher CFE reflects a lower PCF, resulting in less force being transferred to the passenger.

3. Results and Discussion

3.1. Post-Compression Behavior of the Tested Specimens

The post-compression condition of the tested specimens is presented Figure 4. Most configurations demonstrated a favorable progressive folding pattern, characterized by the sequential and uniform formation of folds, indicating stable deformation and efficient energy absorption. The PLA-CF specimens generally exhibited more localized fractures than the PLA+ specimens, suggesting that the inclusion of carbon fibers in PLA-CF may contribute to a more brittle failure response under compressive loading. Specimen T1A0C, however, deviated from this behavior, experiencing brittle fracture instead of folding, which significantly reduced energy absorption. This failure is attributed to its basic cylindrical geometry with a 0° taper angle, a reduced wall thickness of 1 mm, and the inherently brittle nature of PLA-CF. These results indicate that increasing the wall thickness and taper angle may help reduce the occurrence of brittle fracture.

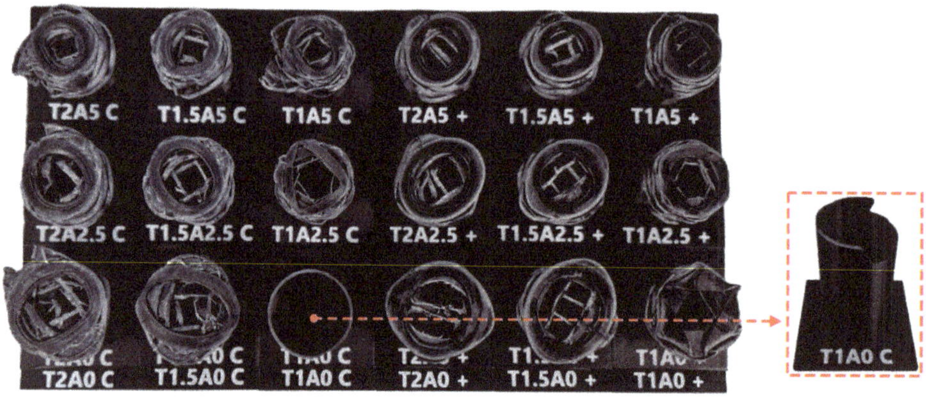

Figure 4. Condition of specimens after compression testing.

3.2. Force-Displacement Curves

The force-displacement curves for the different design configurations are presented in Figure 5. The curves reveal varied behaviors under axial compression, with the majority of configurations showing consistent and uniform plastic deformation, except for T1A0C, which displayed brittle fracture. The *PCF* was observed within the displacement range of 1–5 mm for all configurations. The highest *PCF* was recorded for T2A0+ and T2A0C, at 11.963 kN and 12.155 kN, respectively. In contrast, the lowest *PCF* values were observed in T1A5+ and T1A5C, with forces of 3.934 kN and 4.084 kN. After reaching their peak, all the specimens transitioned into a more stable force region, where the load fluctuated around a mean crush force, indicating continuous energy absorption.

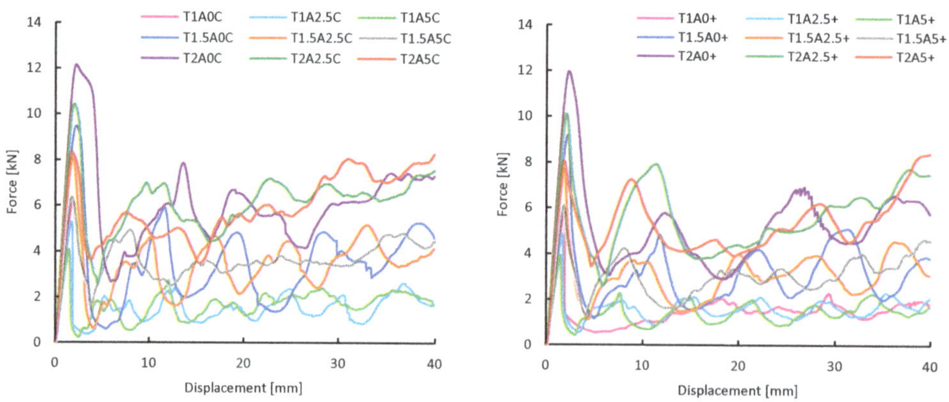

(**a**) (**b**)

Figure 5. Force-displacement curves of different design configurations: (**a**) PLA-CF; (**b**) PLA+.

3.3. Crashworthiness Indicators

The crashworthiness indicators E_T, PCF, MCF, SEA, and CFE for the various design configurations fabricated from PLA-CF and PLA+ are presented in Table 4, including the mass of each specimen. Column charts are utilized to visually present the mass and the crashworthiness indicators, as illustrated in Figure 6. The mass of the specimens clearly depends on both the geometric configuration and the material type. Specifically, the mass increases with wall thickness and decreases with larger taper angles. PLA-CF consistently displays values comparable to or slightly lower than PLA+, which is attributable to its slightly lower density (1.21 g/cm^3) relative to that of PLA+ (1.23 g/cm^3). The highest mass, recorded in T2A0C and T2A0+, is 12.2 g, while the lowest, observed in T1A5C, is 5.1 g.

Table 4. Crashworthiness indicators for various design configurations fabricated from PLA-CF and PLA+.

Case	m [g]	E_T [kJ]	PCF [kN]	MCF [kN]	SEA [kJ/kg]	CFE
T1A0C	6.2	0.007	6.158	0.172	1.112	0.028
T1A2.5C	5.6	0.063	5.268	1.569	11.205	0.298
T1A5C	5.1	0.068	4.084	1.691	13.260	0.414
T1.5A0C	9.3	0.139	9.471	3.466	14.908	0.366
T1.5A2.5C	8.5	0.138	8.087	3.443	16.203	0.426
T1.5A5C	7.6	0.145	6.358	3.631	19.110	0.571
T2A0C	12.2	0.242	12.155	6.040	19.803	0.497
T2A2.5C	11.1	0.243	10.442	6.073	21.884	0.582
T2A5C	10.0	0.235	8.346	5.875	23.500	0.704
T1A0+	6.3	0.059	5.821	1.484	9.421	0.255
T1A2.5+	5.8	0.064	4.855	1.605	11.072	0.331
T1A5+	5.2	0.057	3.934	1.419	10.915	0.361
T1.5A0+	9.3	0.125	9.180	3.135	13.486	0.342
T1.5A2.5+	8.5	0.123	7.728	3.068	14.436	0.397
T1.5A5+	7.7	0.122	6.082	3.042	15.800	0.500
T2A0+	12.2	0.208	11.963	5.191	17.019	0.434
T2A2.5+	11.1	0.222	10.121	5.557	20.026	0.549
T2A5+	10.0	0.211	8.392	5.271	21.082	0.628

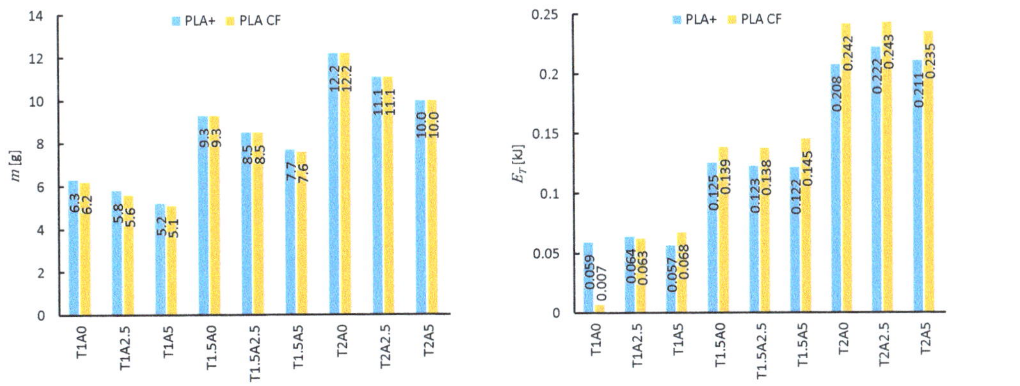

(a) (b)

Figure 6. *Cont.*

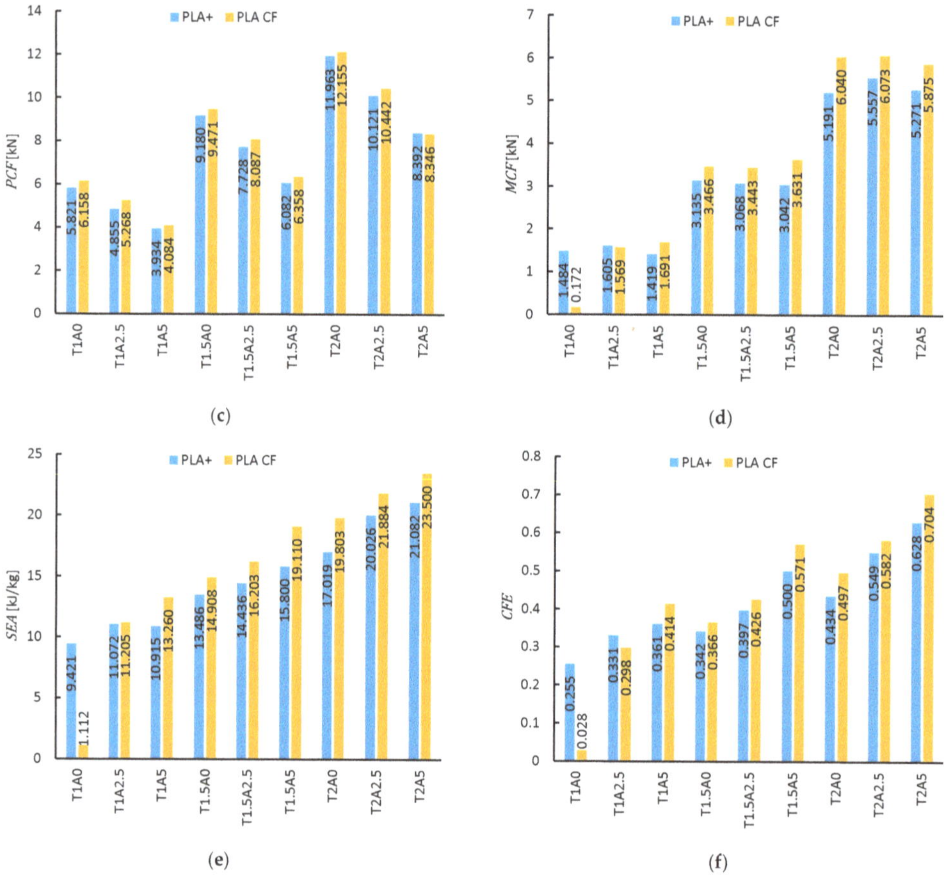

Figure 6. The mass of each specimen (**a**) and the crashworthiness indicators: (**b**) E_T; (**c**) *PCF*; (**d**) *MCF*; (**e**) *SEA*; (**f**) *CFE*.

For E_T, values rise with increased wall thickness and show minimal sensitivity to taper angle variations. PLA-CF generally outperforms PLA+ in E_T, except in configuration T1A0C, where brittle fracture limited the value to 0.01 kJ. The maximum E_T is achieved in T2A2.5C at 0.243 kJ.

PCF is influenced by both geometric parameters, increasing with thickness and decreasing with greater taper angle, a well-established behavior in tapered structures. PLA-CF generally yields a slightly higher *PCF* than PLA+. The highest value is observed in T2A0C at 12.155 kN and the lowest in T1A5+ at 3.934 kN.

MCF follows a trend comparable to E_T, with values rising as thickness increases and showing limited dependence on the taper angle. PLA-CF typically produces a higher *MCF* than PLA+, aside from T1A0C, which experienced brittle failure, reducing the *MCF* to 0.172 kN. The peak value, 6.073 kN, is attained at T2A2.5C.

SEA is responsive to both geometric parameters, increasing with thickness and taper angle. Although E_T remains largely unchanged with the taper angle, the concurrent reduction in mass results in an elevated *SEA*, as it is dependent on both. A higher *SEA* indicates enhanced energy absorption per unit mass, contributing to reduced kinetic energy transfer to occupants. PLA-CF generally exhibits superior *SEA* performance to that of

PLA+. T2A5C achieves the highest SEA at 23.5 kJ/kg, whereas T1A0C, due to brittle failure, shows the lowest value, at 1.112 kJ/kg.

CFE similarly increases with both thickness and taper angle. While MCF remains relatively stable with changing taper angle, the associated decline in PCF improves CFE, as it is defined by the MCF-to-PCF ratio. A higher CFE reflects better load stability during deformation, minimizing the force transferred to occupants. PLA-CF configurations consistently yield higher CFE values than PLA+, with the most favorable outcome observed in T2A5C at 0.704. Conversely, the lowest value of 0.028 occurs in T1A0C due to brittle failure.

4. Taguchi Analysis of Crashworthiness Indicators

In this study, Taguchi analysis was conducted using Minitab statistical software (version 19) to investigate the influence of design parameters on crashworthiness performance. The main effects plots for means provide a clear visualization of how each factor level affects the response variables. These plots, shown in Figure 7, highlight the trends in SEA and CFE across different parameter levels. These trends suggest that the use of PLA-CF over PLA+ and increasing both thickness and taper angle can improve both SEA and CFE.

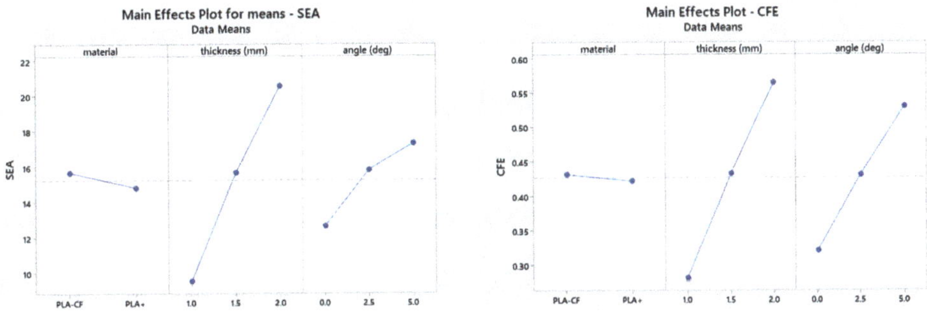

(a) (b)

Figure 7. Main effects plots for means: (**a**) SEA; (**b**) CFE.

The response tables for means corresponding to *SEA* and *CFE* are presented in Table 5. It is observed that the most influential factor is the tube thickness for both *SEA* and *CFE*, whereas the least influential factor is the tube material. The taper angle is the second most influential factor, and its effect is more pronounced on *CFE* than on *SEA*.

Table 5. Response table for the means of *SEA* and *CFE*.

	SEA			*CFE*		
Level	**Material**	**Thickness [mm]**	**Angle [°]**	**Material**	**Thickness [mm]**	**Angle [°]**
1	15.665	9.498	12.625	0.4317	0.2810	0.3202
2	14.806	15.657	15.804	0.4218	0.4336	0.4303
3		20.552	17.278		0.5656	0.5296
Delta	0.859	11.055	4.653	0.0099	0.2846	0.2094
Rank	3	1	2	3	1	2

5. ANOVA of Crashworthiness Indicators

A statistical analysis of variance (ANOVA) was performed using Minitab statistical software (version 19) to quantitatively assess the effects of the geometric and material parameters on crash performance. The factors included the continuous variables (thickness and taper angle) as well as the categorical variable (material type). The ANOVA results are listed in Tables 6 and 7. Thickness is the most influential factor, accounting for 73.18% of the variation in SEA and 58.19% in CFE, confirming that increasing wall thickness substantially improves crash performance. Taper angle contributes 13.49% of the variation in SEA and 31.49% in CFE, highlighting the importance of geometric design in crashworthiness optimization. Material type accounts for smaller contributions of 0.66% in SEA and 0.11% in CFE. The residual error accounts for 12.67% and 10.22% of the variation in SEA and CFE, respectively. For both SEA and CFE, the p-values corresponding to thickness and angle indicate that the ANOVA results related to these parameters are statistically significant, whereas the p-values corresponding to material show that the ANOVA results related to material are statistically insignificant.

Table 6. ANOVA results for SEA.

Source of Variation	DF	Contribution	Adj SS	Adj MS	p-Value
Material	1	0.66%	3.318	3.318	0.44478
Thickness [mm]	2	73.18%	368.223	184.111	0.00001
Angle [°]	2	13.49%	67.868	33.934	0.01293
Error	12	12.67%	63.771	5.314	
Total	17	100%			

Table 7. ANOVA results for CFE.

Source of Variation	DF	Contribution	Adj SS	Adj MS	p-Value
Material	1	0.11%	0.000441	0.000441	0.73095
Thickness [mm]	2	58.19%	0.243347	0.121673	0.00001
Angle [°]	2	31.49%	0.131694	0.065847	0.00022
Error	12	10.22%	0.042732	0.003561	
Total	17	100%			

6. Regression Analysis

The experiment involved varying three parameters: thickness, taper angle, and filament material. These parameters were used to develop regression models for predicting output responses, specifically for SEA and CFE. To model these responses, a second-order polynomial response surface (PRS) approach was used. The general form of the second-order polynomial model is expressed as follows [42]:

$$\hat{f} = b_0 + \sum_{i=1}^{L} b_i x_i + \sum_{i=1}^{L} b_{ii} x_i^2 + \sum_{i=1}^{L-1} \sum_{j=i+1}^{L} b_{ij} x_i x_j \tag{5}$$

where $\hat{f}$ represents an approximation of the actual response function f (SEA or CFE), x consists of L variables (thickness T and taper angle A), and the coefficients b_0, b_i, b_{ii}, and b_{ij} are the regression coefficients determined using the least-squares technique. For each filament material, two separate regression models were developed, one for SEA and one for CFE. MATLAB (version R2013b) functions polyfitn and polyvaln were utilized for the polynomial fitting and evaluation of the model predictions, respectively. Specimen T1A0C was excluded from the training points due to early specimen failure, potentially leading to incorrect model predictions. The regression models for PLA-CF and PLA+ are presented below:

$$SEA_{PLA-CF} = -1.045579T^2 - 0.068229TA + 13.807217T + 0.032447A^2 + 0.735893A - 3.539760 \tag{6}$$

$$CFE_{PLA-CF} = -0.001494T^2 + 0.004472TA + 0.273776T + 0.004555A^2 + 0.010306A - 0.043008 \tag{7}$$

$$SEA_{PLA+} = 1.393394T^2 + 0.513630TA + 3.442063T - 0.089180A^2 + 0.200288A + 4.778018 \tag{8}$$

$$SEA_{PLA+} = 1.393394T^2 + 0.513630TA + 3.442063T - 0.089180A^2 + 0.200288A + 4.778018 \quad (9)$$

To evaluate the accuracy of the generated models, two numerical estimators, namely R-squared (R^2) and Maximum Absolute Percentage Error ($MAPE$), were employed to validate the models, as outlined in Equations (10) and (11) [36]. The resulting values are provided in Table 8.

$$R^2 = 1 - \frac{\sum_{i=1}^{n} (y_i - \hat{y}_i)^2}{\sum_{i=1}^{n} (y_i - \overline{y}_i)^2} \quad (10)$$

$$MAPE = max\left(\frac{|\hat{y}_i - y_i|}{y_i}\right) \times 100 \quad (11)$$

where n is the total number of data points, y_i is the actual value, $\hat{y}_i$ is the predicted value, and $\overline{y}_i$ is the mean of the actual values.

Table 8. Error assessment of the regression models for PLA-CF and PLA+.

Material	Crashworthiness Indicator	R^2	$MAPE$
PLA-CF	SEA	0.99671	2.48%
	CFE	0.99804	2.23%
PLA+	SEA	0.99503	3.53%
	CFE	0.99189	4.97%

The 3D contour plots of the SEA and CFE models for both PLA-CF and PLA+ are illustrated in Figure 8, providing a detailed visualization of their predictive performance. In all the plots, the regression models accurately interpolate the training data. It is observed that both SEA and CFE increase with greater thickness and taper angle, with PLA-CF consistently outperforming PLA+. Higher SEA and CFE values indicate enhanced crashworthiness by improving energy absorption and reducing the force transmitted to occupants. These trends suggest that increasing both thickness and taper angle can lead to improved crash performance. Based on this observation, three additional design cases were proposed, as shown in Figure 9a: T2A7.5C, T2.5A7.5C, and T3A7.5C.

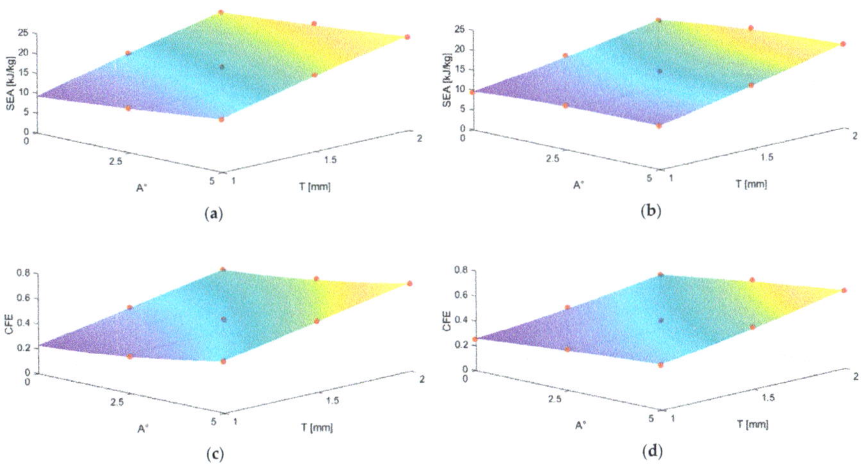

Figure 8. Regression-based response surfaces, with training points indicated by red dots: (**a**) SEA PLA-CF; (**b**) SEA PLA+; (**c**) CFE PLA-CF; (**d**) CFE PLA+.

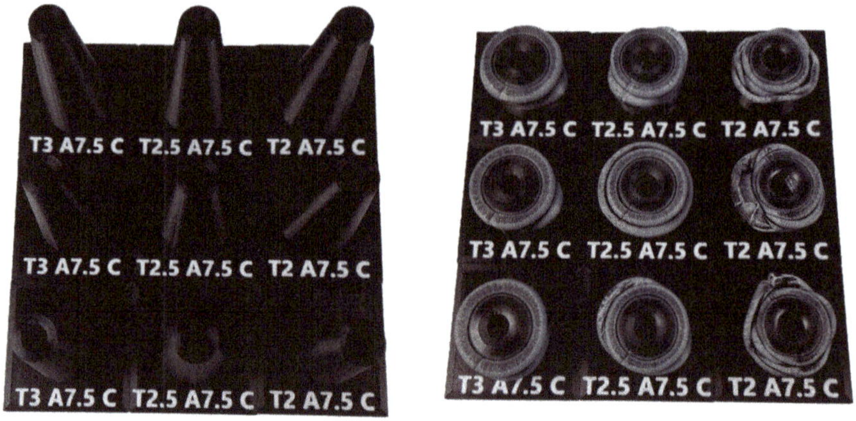

(a) (b)

Figure 9. Proposed design configurations: (**a**) before compression; (**b**) after compression.

The condition of the T2A7.5C, T2.5A7.5C, and T3A7.5C specimens after the compression tests is shown in Figure 9b. Each configuration was tested three times, and the mean values, along with the corresponding standard deviations of the crashworthiness indicators and specimen mass, are presented in Table 9. In addition, the corresponding force-displacement curves for all three replicates of each configuration are illustrated in Figure 10. Among the configurations, T3A7.5C consistently demonstrated the highest performance, achieving an average SEA of 32.086 $\pm$ 0.190 kJ/kg and an average CFE of 0.745 $\pm$ 0.034. These results confirm the superior crashworthiness of the T3A7.5C design and validate the positive influence of increased thickness on both SEA and CFE. The low standard deviations reported indicate good repeatability of the experimental results, supporting the reliability of the regression model-guided design strategy.

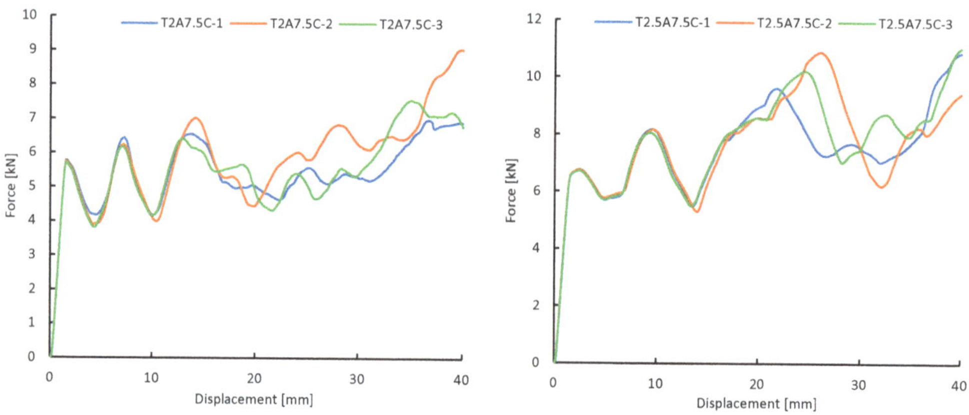

(a) (b)

Figure 10. *Cont.*

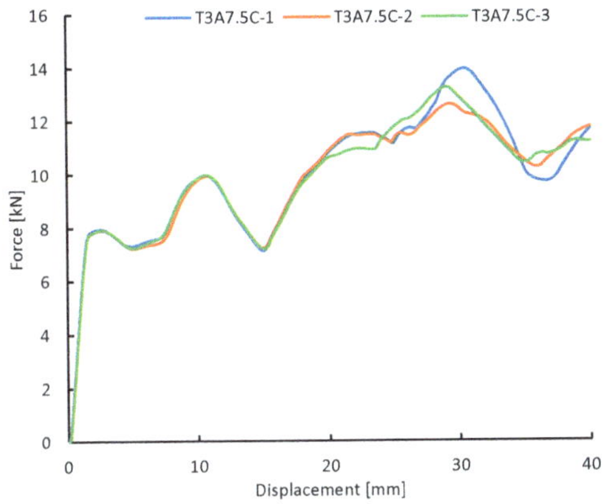

(c)

Figure 10. Force-displacement of the three test replicates for each proposed design configuration: (**a**) T2A7.5C; (**b**) T2.5A7.5C; (**c**) T3A7.5C.

Table 9. The mean and standard deviations obtained through three test replicates for the proposed design configurations.

Case	*m* [g]	E_T [kJ]	*PCF* [kN]	*MCF* [kN]	*SEA* [kJ/kg]	*CFE*
T2A7.5C	8.7 ± 0.0	0.224 ± 0.009	7.860 ± 1.065	5.588 ± 0.227	25.691 ± 1.042	0.717 ± 0.065
T2.5A7.5C	10.6 ± 0.0	0.305 ± 0.004	10.916 ± 0.090	7.632 ± 0.108	28.800 ± 0.409	0.699 ± 0.005
T3A7.5C	12.3 ± 0.0	0.395 ± 0.002	13.272 ± 0.674	9.866 ± 0.058	32.086 ± 0.190	0.745 ± 0.034

A comparative evaluation was conducted between the optimal crash box developed in the present study and the optimal design configurations reported in the recent literature concerning FDM-fabricated tubular crash boxes, as depicted in Figure 11. The figure illustrates the correlation between *SEA* and *CFE* for structures manufactured using various materials, including PLA, PLA+, PLA-LW, PA-CF, and TPU. Among the compared designs, the T3A7.5C specimen exhibits the highest *SEA* while maintaining a substantial *CFE* relative to the crash boxes reported in previous studies. These findings highlight the strong potential of PLA-CF, particularly in tapered geometries, for high-performance crash box applications.

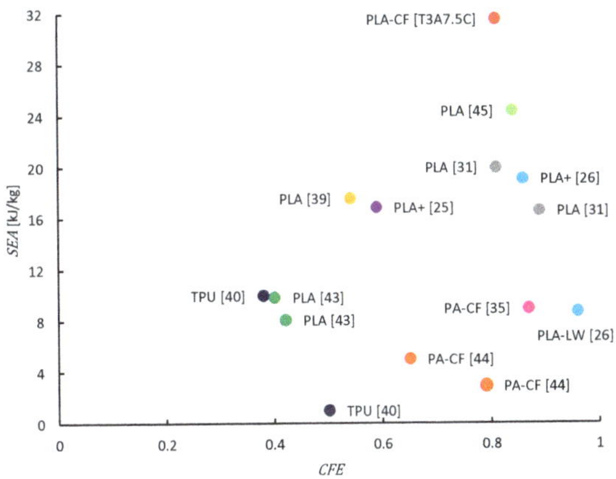

Figure 11. Comparison of *SEA* and *CFE* values for tubular crash boxes manufactured using the FDM process in recent studies [25,26,31,35,39,40,43–45].

7. Limitations and Practical Implications

Despite the promising energy-absorption characteristics demonstrated by the tapered FDM-printed PLA+ and PLA-CF structures, several limitations restrict their direct application in high-performance or safety-critical automotive environments. PLA-based materials, including PLA-CF, exhibit inherent drawbacks such as relatively low thermal stability, limited impact resistance at elevated temperatures, and susceptibility to moisture absorption. These factors compromise their long-term durability and reliability under real-world operating conditions.

Nevertheless, PLA+ and PLA-CF remain valuable for rapid prototyping and early-stage design validation. Their low cost and ease of fabrication make them suitable for iterative development cycles, the functional testing of geometric concepts, and applications in lightweight, low-speed vehicles, robotics, and other products where crash energy demands are less strict.

Future research should focus on translating the optimized geometries identified in this study to higher-performance additive manufacturing materials, including nylon, carbon fiber-reinforced nylon, or metal-based AM processes. Such approaches have the potential to enhance thermal resistance, mechanical performance, and overall structural integration, thereby improving their feasibility for real-world operational conditions.

8. Conclusions

In this study, the crashworthiness performance of 3D-printed tapered-tube crash boxes was investigated. The specimens were fabricated using Fused Deposition Modeling (FDM) with two advanced filament materials: PLA-CF and PLA+. A series of nine geometric configurations were designed by varying wall thicknesses and taper angles, and each configuration was printed using both materials to comprehensively evaluate their structural response under quasi-static axial compression. The results demonstrate that greater wall thickness and larger taper angles significantly improve energy-absorption characteristics. Across all configurations, PLA-CF consistently outperformed PLA+ in both SEA and CFE, indicating its suitability for energy-dissipating applications. ANOVA confirmed that thickness, taper angle, and material type significantly affect SEA and CFE. Thickness was the most influential factor, accounting for 73.18% of the variation in SEA and 58.19% in CFE, followed by taper angle, with contributions of 13.49% and 31.49%, respectively. Material type had smaller but meaningful impacts, contributing 0.66% to SEA and 0.11% to CFE. Based on the experimental dataset, second-order polynomial regression models were developed to predict SEA and CFE, yielding a maximum absolute percentage error of 4.97%. These models were further utilized to guide the design of new configurations. The optimal design, defined by a thickness of 3 mm and a taper angle of 7.5 degrees using PLA-CF, achieved an SEA of 32.086 ± 0.190 kJ/kg and a CFE of 0.745 ± 0.034. The findings highlight the effectiveness of tapered-tube designs and the use of PLA-CF material in improving crashworthiness performance.

Author Contributions: Conceptualization, A.S. and M.A.G.; methodology, A.S., E.A. and M.A.G.; formal analysis, O.S.E., A.S., E.A. and M.A.G.; investigation, O.S.E., A.S., E.A. and M.A.G.; resources, A.S. and M.A.G.; data curation, H.A., A.S. and M.A.G.; writing—original draft preparation, O.S.E., H.A., A.A. and Y.A.; writing—review and editing, A.S., E.A. and M.A.G.; visualization, A.S., E.A. and M.A.G.; supervision, A.S. and M.A.G.; project administration, M.A.G. All authors have read and agreed to the published version of the manuscript.

Funding: This research received no external funding.

Data Availability Statement: The data presented in this study are available on request from the corresponding author.

Conflicts of Interest: The authors declare no conflicts of interest.

References

1. Omar, M.A. *The Automotive Body Manufacturing Systems and Processes*; John Wiley and Sons: Hoboken, NJ, USA, 2011; ISBN 9780470976333.
2. Bandinelli, F.; Ciampaglia, A.; Ciardiello, R.; Peroni, L.; Peroni, M.; Scapin, M. High Strain Rate Testing of Carbon-Epoxy Laminate Crash Boxes Filled with Polymeric Cellular 3D-Printed Cores. *Int. J. Impact Eng.* **2025**, *205*, 105401. [CrossRef]
3. Chen, J.; Li, E.; Liu, W.; Mao, Y.; Hou, S. Crashworthiness Analysis of Novel Cactus-Inspired Multi-Cell Structures under Axial Crushing. *Int. J. Mech. Sci.* **2024**, *268*, 109053. [CrossRef]
4. Hanssen, A.G.; Langseth, M.; Hopperstad, O.S. Static and Dynamic Crushing of Circular Aluminum Extrusions with Aluminum Foam Filler. *Int. J. Impact Eng.* **2000**, *24*, 475–507. [CrossRef]
5. Hanssen, A.G.; Langseth, M.; Hopperstad, O.S. Static and Dynamic Crushing of Square Aluminum Extrusions with Aluminum Foam Filler. *Int. J. Impact Eng.* **2000**, *24*, 347–383. [CrossRef]
6. Wesselmecking, S.; Kreins, M.; Dahmen, M.; Bleck, W. Material Oriented Crash-Box Design—Combining Structural and Material Design to Improve Specific Energy Absorption. *Mater. Des.* **2022**, *213*, 110357. [CrossRef]
7. Kaczyński, P.; Makuła, P. Constitutive Strength Model for Spot Joints of Thin-Walled Energy-Absorbing Elements. *Eng. Struct.* **2023**, *292*, 116592. [CrossRef]
8. Aktaş, C.; Acar, E.; Güler, M.A.; Altın, M. An Investigation of the Crashworthiness Performance and Optimization of Tetra-Chiral and Reentrant Crash Boxes. *Mech. Based Des. Struct. Mach.* **2023**, *51*, 6881–6904. [CrossRef]
9. Ma, J.; Dai, H.; Shi, M.; Yuan, L.; Chen, Y.; You, Z. Quasi-Static Axial Crushing of Hexagonal Origami Crash Boxes as Energy Absorption Devices. *Mech. Sci.* **2019**, *10*, 133–143. [CrossRef]
10. Yuan, L.; Shi, H.; Ma, J.; You, Z. Quasi-Static Impact of Origami Crash Boxes with Various Profiles. *Thin-Walled Struct.* **2019**, *141*, 435–446. [CrossRef]
11. Hull, C.W.; Arcadia, C. Apparatus for Production of Three Dimensional Objects by Stereolithography. US4575330A, 11 March 1986.
12. Li, Q.; Wang, W.; Tan, H.; Long, X.; Wang, F.; Hu, L. Energy Absorption Characteristics of Modular Assembly Structures under Quasi-Static Compression Load. *Compos. Struct.* **2024**, *342*, 118260. [CrossRef]
13. Wang, Y.J.; Zhang, Z.J.; Xue, X.W.; Zhou, J.; Song, Z.X. Axial and Lateral Crushing Performance of Plate-Lattice Filled Square Sandwich Tubes. *Compos. Struct.* **2021**, *274*, 114404. [CrossRef]
14. Tao, Y.; Li, W.; Wei, K.; Duan, S.; Wen, W.; Chen, L.; Pei, Y.; Fang, D. Mechanical Properties and Energy Absorption of 3D Printed Square Hierarchical Honeycombs under In-Plane Axial Compression. *Compos. Part B Eng.* **2019**, *176*, 107219. [CrossRef]
15. Zhang, X.; Xie, J.; Chen, J.; Okabe, Y.; Pan, L.; Xu, M. The Beetle Elytron Plate: A Lightweight, High-Strength and Buffering Functional-Structural Bionic Material. *Sci. Rep.* **2017**, *7*, 4440. [CrossRef]
16. Xu, W.; Wang, C.; Liu, B.; Jia, S. Crushing Responses and Energy Absorption of Bionic Inspired Corrugated Honeycombs. *Int. J. Impact Eng.* **2023**, *179*, 104641. [CrossRef]
17. Yang, K.; Li, Z.; Ge, D. Quasi-Static and Dynamic out-of-Plane Crashworthiness of 3D Curved-Walled Mixed-Phase Honeycombs. *Thin-Walled Struct.* **2023**, *182*, 110305. [CrossRef]
18. Tang, Y.; Li, Q.; Miao, X.; Chen, B.; Yin, L.; Liu, X.; Gu, C. Design Novel Origami Structures for Energy Absorption under Compressive Load. *Acta Mech. Sin. Xuebao* **2024**, *40*, 423271. [CrossRef]
19. Zeng, Y.; Du, X.; Yao, H.; Li, P.; Dong, P.; Chen, J. An Nylon Lattice Structure with Improved Mechanical Property and Energy Absorption Capability. *Compos. Part C Open Access* **2022**, *8*, 100285. [CrossRef]
20. Peng, X.; Liu, G.; Li, J.; Wu, H.; Jia, W.; Jiang, S. Compression Property and Energy Absorption Capacity of 4D-Printed Deformable Honeycomb Structure. *Compos. Struct.* **2023**, *325*, 117591. [CrossRef]
21. Raj, R.; Jiyalal Prajapati, M.; Tsai, J.T.; Kumar, A.; Jeng, J.Y. Design and Additive Manufacturing of Novel Hybrid Lattice Metamaterial for Enhanced Energy Absorption and Structural Stability. *Mater. Des.* **2024**, *245*, 113268. [CrossRef]
22. Shen, W.; Zhang, Z.; Okudan Kremer, G.E.; Qin, H. Origami-Inspired Infill Pattern for Additive Manufacturing. *Manuf. Lett.* **2022**, *33*, 516–520. [CrossRef]
23. Capasso, I.; Andreacola, F.R.; Brando, G. Additive Manufacturing of Metal Materials for Construction Engineering: An Overview on Technologies and Applications. *Metals* **2024**, *14*, 1033. [CrossRef]
24. Zhou, J.; Liu, H.; Dear, J.P.; Falzon, B.G.; Kazancı, Z. Comparison of Different Quasi-Static Loading Conditions of Additively Manufactured Composite Hexagonal and Auxetic Cellular Structures. *Int. J. Mech. Sci.* **2023**, *244*, 108054. [CrossRef]

25. Tunay, M.; Bardakci, A. A Study of Crashworthiness Performance in Thin-Walled Multi-Cell Tubes 3D-Printed from Different Polymers. *J. Appl. Polym. Sci.* **2024**, *141*, e56287. [CrossRef]
26. Hidayat, D.; Istiyanto, J.; Sumarsono, D.A.; Kurniawan, F.; Ardiansyah, R.; Wandono, F.A.; Nugroho, A. Investigation on the Crashworthiness Performance of Thin-Walled Multi-Cell PLA 3D-Printed Tubes: A Multi-Parameter Analysis. *Designs* **2023**, *7*, 108. [CrossRef]
27. Zhang, D.; Li, M.; Qiu, N.; Yang, J.; Wu, C.; Steven, G.; Li, Q.; Fang, J. 4D-Printed Reusable Metamaterial via Shape Memory Effect for Energy Dissipation. *Int. J. Mech. Sci.* **2024**, *275*, 109309. [CrossRef]
28. Isaac, C.W.; Sokołowski, A.; Duddeck, F.; Adamiak, M.; Pakieła, W.; Aremu, A. Mechanical Characterisation and Crashworthiness Performance of Additively Manufactured Polymer-Based Honeycomb Structures under in-Plane Quasi-Static Loading. *Virtual Phys. Prototyp.* **2023**, *18*, e2273296. [CrossRef]
29. Wang, J.; Liu, Y.; Wang, K.; Yao, S.; Peng, Y.; Rao, Y.; Ahzi, S. Progressive Collapse Behaviors and Mechanisms of 3D Printed Thin-Walled Composite Structures under Multi-Conditional Loading. *Thin-Walled Struct.* **2022**, *171*, 108810. [CrossRef]
30. Sun, G.; Wang, J.; Wang, K.; Baghani, M.; Peng, Y.; Rao, Y. Repeatable Compressive Functionality of 3D Printed Shape-Memory Thin-Walled Corrugated Structures. *Int. J. Mech. Sci.* **2023**, *257*, 108552. [CrossRef]
31. Cetin, E. Energy Absorption of Thin-Walled Multi-Cell Tubes with DNA-Inspired Helical Ribs under Quasi-Static Axial Loading. *J. Braz. Soc. Mech. Sci. Eng.* **2024**, *46*, 607. [CrossRef]
32. Ha, N.S.; Pham, T.M.; Tran, T.T.; Hao, H.; Lu, G. Mechanical Properties and Energy Absorption of Bio-Inspired Hierarchical Circular Honeycomb. *Compos. Part B Eng.* **2022**, *236*, 109818. [CrossRef]
33. Wang, K.; Liu, Y.; Wang, J.; Xiang, J.; Yao, S.; Peng, Y. On Crashworthiness Behaviors of 3D Printed Multi-Cell Filled Thin-Walled Structures. *Eng. Struct.* **2022**, *254*, 113907. [CrossRef]
34. Liu, Y.; Wang, J.; Tan, Q.; Gao, H.; Wang, K.; Yao, S.; Peng, Y. On Multi-Stage Deformation and Gradual Energy Absorption of 3D Printed Multi-Cell Tubes with Varying Cross-Section. *Eng. Struct.* **2024**, *319*, 118839. [CrossRef]
35. Liu, Y.; Tan, Q.; Lin, H.; Wang, J.; Wang, K.; Peng, Y.; Yao, S. Integrated Design and Additive Manufacturing of Lattice-Filled Multi-Cell Tubes. *Compos. Sci. Technol.* **2023**, *243*, 110252. [CrossRef]
36. Song, X.; Sun, G.; Li, G.; Gao, W.; Li, Q. Crashworthiness Optimization of Foam-Filled Tapered Thin-Walled Structure Using Multiple Surrogate Models. *Struct. Multidiscip. Optim.* **2013**, *47*, 221–231. [CrossRef]
37. Mathiazhagan, N.; Palaniyappan, S.; Sivakumar, N. kumar Effect of Fused Filament Fabrication Parameters on Crashworthiness Studies of Hydroxyapatite Particle Reinforced PLA Composite Thin-Walled Tubes. *J. Mech. Behav. Biomed. Mater.* **2023**, *138*, 105611. [CrossRef] [PubMed]
38. Li, J.; Qu, M.; Jiang, Z. Multi-Objective Optimization for 3D Printed Origami Crash Box Cell Based on Artificial Neural Networks and NSGA-II. *J. Mater. Process. Des.* **2023**, *7*, 1–13. [CrossRef]
39. Hidayat, D.; Istiyanto, J.; Sumarsono, D.A.; Kurniawan, F.; Abdurohman, K. Crashworthiness Characteristics of Thin-Walled Multi-Cell Structures via the Application of Annealing Process: An Experimental Investigation. *Evergreen* **2024**, *11*, 1892–1900. [CrossRef]
40. Wang, K.; Sun, G.; Wang, J.; Yao, S.; Baghani, M.; Peng, Y. Reversible Energy Absorbing Behaviors of Shape-Memory Thin-Walled Structures. *Eng. Struct.* **2023**, *279*, 115626. [CrossRef]
41. Saber, A.; Güler, M.A.; Altin, M.; Acar, E. Bio-Inspired Thin-Walled Energy Absorber Adapted from the Xylem Structure for Enhanced Vehicle Safety. *J. Braz. Soc. Mech. Sci. Eng.* **2024**, *46*, 613. [CrossRef]
42. Acar, E.; Rais-Rohani, M. Ensemble of Metamodels with Optimized Weight Factors. *Struct. Multidiscip. Optim.* **2009**, *37*, 279–294. [CrossRef]
43. Wirawan, W.A.; Junipitoyo, B.; Putro, S.H.S.; Sabitah, A.; Suudy, A.H.; Ridwan, R.; Choiron, M.A. Collapse Behavior and Energy Absorption Characteristics of Design Multi-Cell Thin Wall Structure 3D-Printed Under Quasi Statistic Loads. *Automot. Exp.* **2024**, *7*, 149–160. [CrossRef]
44. Wang, K.; Tan, Q.; Wang, J.; Liu, Y.; Zhai, Z.; Yao, S.; Peng, Y. A Low-Cost Granular-Medium Hot Quasi-Isostatic Pressing Method for Enhancing Compressive Properties of 3D Printed Structures. *J. Manuf. Process.* **2024**, *115*, 441–451. [CrossRef]
45. Awd Allah, M.M.; Abdel-Aziem, W.; Abd El-baky, M.A. Collapse Behavior and Energy Absorbing Characteristics of 3D-Printed Tubes with Different Infill Pattern Structures: An Experimental Study. *Fibers Polym.* **2023**, *24*, 2609–2622. [CrossRef]

designs

MDPI

Article

A Flake Powder Metallurgy Approach for Fabricating Al/CNT Composites: Combining Dual-Matrix and Shift-Speed Ball Milling to Optimize Mechanical Properties

Hamed Rezvanpour and Alberto Vergnano *

Enzo Ferrari Department of Engineering, University of Modena and Reggio Emilia, Via P. Vivarelli 10, 41125 Modena, Italy; hamed.rezvanpour@unimore.it
* Correspondence: alberto.vergnano@unimore.it

Abstract: This study presents a novel flake powder metallurgy approach for fabricating Al/CNT composites, combining the dual-matrix (DM) method with shift-speed ball milling (SSBM) to optimize mechanical performance. Samples prepared via DM-SSBM were systematically compared to those produced by conventional high-speed ball milling (HSBM), single-stage SSBM, and dual-matrix (DM) routes. Tensile testing revealed that the DM_1MR_{50}-SSBM composite achieved a superior balance of strength and ductility, with an ultimate tensile strength of ~267 MPa, elongation of ~9.9%, and the highest energy absorption capacity (~23.4 MJ/m^3) among all tested samples. In contrast, the HSBM sample, while achieving the highest tensile strength (~328 MPa), exhibited limited elongation (~4.7%), resulting in lower overall toughness. The enhanced mechanical response of the DM-SSBM composites is attributed to improved CNT dispersion, refined cold-welding interfaces, and pure Al matrix softness, which together facilitate superior load transfer and hinder crack propagation under tensile stress. In the final consolidated state, aluminum forms a continuous matrix embedding the CNTs, justifying the use of the term "aluminum matrix" to describe the composite structure. These findings highlight the DM-SSBM approach as a promising method for developing lightweight, high-toughness aluminum composites suitable for energy-absorbing structural applications.

Keywords: composites; CNT; aluminum; nanocomposite; powder metallurgy

1. Introduction

The development of lightweight, high-performance structural materials is highly driven by the increasing demands of industries such as aerospace, automotive, and defense, where components must deliver high strength, energy absorption capability, and thermal stability under extreme conditions [1,2]. In this context, metal matrix composites reinforced with nanocarbon materials such as carbon nanotubes (CNTs) have attracted significant attention as promising structural and functional materials, due to the CNTs' remarkable thermal, electrical, and mechanical properties [3–5]. CNTs are the strongest artificial material, with an ultimate tensile strength of 1 TPa [6]. As well as the tensile strength, the high aspect ratio and low density make these materials an ideal reinforcement for ceramics [7], polymers [8–10], and metal matrix composites [11–13]. Among MMC/CNTs, Al/CNTs [14–31], Cu/CNT [32–36], and Mg/CNT [37–40] are most popular due to their lightweight and enhanced mechanical properties. Aluminum was selected in this study as the base matrix due to its low density, high specific strength, and excellent formability, which make it a key material in the automotive, aerospace, and energy sectors. Aluminum was selected as the base metallic phase, which, after processing, forms a continuous matrix embedding the CNTs. While initially used as discrete powder particles, aluminum becomes the dominant matrix phase after consolidation.

Fabricating CNT-based nanocomposites with desired properties depends mainly on (a) the fabrication method (mainly the dispersion method) and (b) the quality of raw materials [15]. The main challenge in fabricating an MMC/CNT is to disperse the CNTs in metal powder because CNTs tend to agglomerate due to the poor wettability between the matrix and the tubes, as well as the strong van der Waals force between the tubes [15,41,42]. Powder diameter, particle size, and CNT length are also important in the dispersion quality. F. Rikhtegar et al. [16,43] found that the perfect mesh size for Al matrix to fabricate a ductile Al/CNT nanocomposite is 20 μm. They also reported that the length of the CNTs has to be close enough to the Al mesh size in order to fabricate a ductile nanocomposite.

Over the past few years, aluminum matrix reinforced by CNTs (Al/CNT) has been fabricated by different techniques, namely powder metallurgy, melt processing, thermal spraying, and other novel methods such as friction stir processing and nanoscale dispersion [5]. The powder metallurgy routes, such as mechanical alloying (MA) and spark plasma sintering (SPS), are the best approaches to fabricate Al/CNTs, because of their high effectiveness, high parameter controllability, and also their uniform CNT dispersion [15]. On the other hand, conventional melt-processing methods, due to some constraints such as excessive damage to CNT structure, carbide phase formation, poor wetting properties of CNTs, and a high amount of energy needed [15,44], are not as effective as the powder metallurgy methods. For dispersing the CNTs in the Al matrix using powder metallurgy and mechanical alloying (MA), shift-speed ball milling (SSBM) [14,45], and the dual-matrix (DM) [15] methods demonstrated high ductility, uniform CNT dispersion, and minimum CNT damages, which result in low carbide formation. The SSBM contains a long period of low-speed ball milling (LSBM) to create flake-shaped Al particles, resulting in a uniform CNT dispersion without any excessive damage to CNT structure and a short period of high-speed ball milling (HSBM), which leads to fabricating cold-welded and perfect bonded Al/CNT microstructures [45]. The DM method, which includes adding a soft Al phase after dispersing the CNTs in the initial Al phase, results in perfect formability and better ductility [15]. The milling parameters, such as milling speed, milling time, ball-to-powder ratio (BPR), and milling atmosphere, are significantly affecting the final mechanical properties of the nanocomposite based on the previous reports [28,46–48]. Accordingly, controlling these parameters is a vital part of fabricating an Al/CNT nanocomposite.

To obtain better dispersion and higher density, a deformation step is required to break down the remaining CNT clusters and align them with the deformation direction in order to improve the load transfer. Hot extrusion, hot and cold rolling, and equal channel angular extrusion are the most common deformation processes used by the researchers [5]. In the hot extrusion process, the Extrusion Ratio (ER) is the main parameter that influences the mechanical properties of the final sample. The ER = 25 is considered to be the optimum ratio to obtain the best mechanical properties when the CNT weight percentage is under 2% [48]. Few researchers have investigated the Al/CNT nanocomposites fabricated by hot/cold rolling [49]. They demonstrated that hot rolling following powder mixing or planetary milling facilitated the uniform distribution and partial alignment of CNTs within aluminum matrices, contributing to increased tensile strength and structural cohesion.

Despite the progress made using dual-matrix (DM) and high-speed ball milling (HSBM) techniques, these approaches remain limited by their trade-offs: DM enhances ductility but often lacks sufficient strength due to inadequate reinforcement dispersion, while HSBM improves strength but frequently damages CNTs and compromises ductility [1]. Recently, shift-speed ball milling (SSBM) has emerged as a promising alternative to enhance dispersion while reducing CNT breakage through controlled variation in shear intensity [2]; however, it has not yet been integrated with dual-matrix strategies to simultaneously optimize strength and ductility. To the best of the research team's knowledge, no prior study has combined the DM and SSBM approaches into a unified process. This work addresses this gap by proposing a flake powder metallurgy method that merges the gradual blending effect of DM with the structural preservation advantages of SSBM to fabricate Al–CNT composites with balanced mechanical performance.

Therefore, a novel and mechanically efficient flake powder metallurgy method, which has combined the SSBM and DM methods, was developed, aiming to fabricate an Al/CNT nanocomposite with a balanced strength and ductility to boost the potential applications. This innovative method is created based on the previous literature [3,15,45,47,49]. While the formation of reinforcement-rich and reinforcement-lean regions has been explored in previous dual-matrix (DM) studies, the novelty of this work is in the integration of this concept with shift-speed ball milling (SSBM) to achieve simultaneous control over reinforcement distribution and mechanical properties. This combined strategy, implemented within a single-cycle flake powder metallurgy process, enables not only the spatial design of the matrix but also the preservation of CNT structural integrity—a limitation in DM or high-speed ball-milling methods. The combination of different reinforcement areas and carefully controlled CNT processing improves the microstructure, leading to better strength and ductility.

2. Materials and Methods

2.1. Raw Materials

In this study, aluminum powder 6061 with a 20 μm mesh size (HNYY powder Co., Ltd. Henan, China) was used in combination with multi-walled carbon nanotubes (MWCNTs) with an internal diameter of 5–10 nm and 10–30 μm in length (DYNANO Co., Ltd. China). Table 1 summarizes the key attributes of nanotubes.

Table 1. MWCNT main attributes.

Single-Walled Carbon Nanotubes	
Outside Diameter	20–30 nm
Inside Diameter	5–10 nm
Purity	>95 wt%
Length	10–30 micron
Electrical Conductivity	>100 s/cm
Tap Density	0.28 g/cm^3
True Density	~2.1 g/cm^3
Young's Modulus (GPa)	1200
Tensile Strength (Gpa)	150

MWCNTs are used over SWCNTs for two main reasons: firstly, SWCNTs are mostly used in electrical applications because of their unique electronic properties. Secondly, SWCNTs are more challenging to produce in high purity; consequently, they are often way more expensive than MWCNTs, which limits their application, especially in the mechanical and material engineering fields. Figure 1 shows the SEM images of the Al and CNT powder.

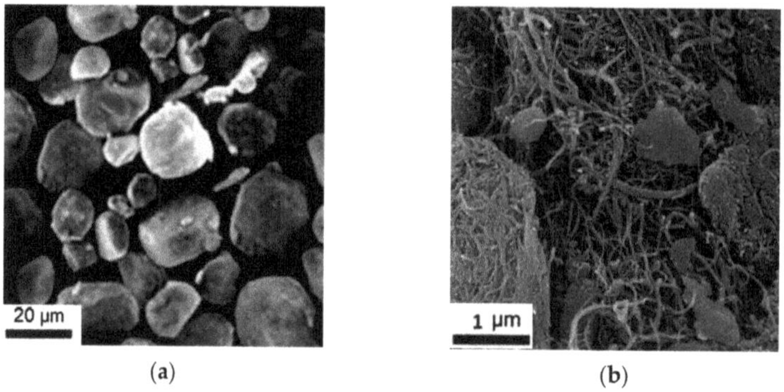

Figure 1. SEM images of (**a**) aluminum powder, (**b**) CNTs.

2.2. *Al\CNT Powder Preparation*

Figure 2 shows the dispersion procedure flowchart in the DM-SSBM method. Two main steps were needed to fabricate the AL/CNT powders: (1) The SSBM phase contains LSBM (175 rpm) and HSBM (400 rpm), and (2) the DM phase contains adding unmilled AL particles to the powder and milling for 1 h in HSBM (400 rpm). Three more samples using conventional HSBM, SSBM, and DM methods were synthesized to compare the final mechanical properties with the DM-SSBM method.

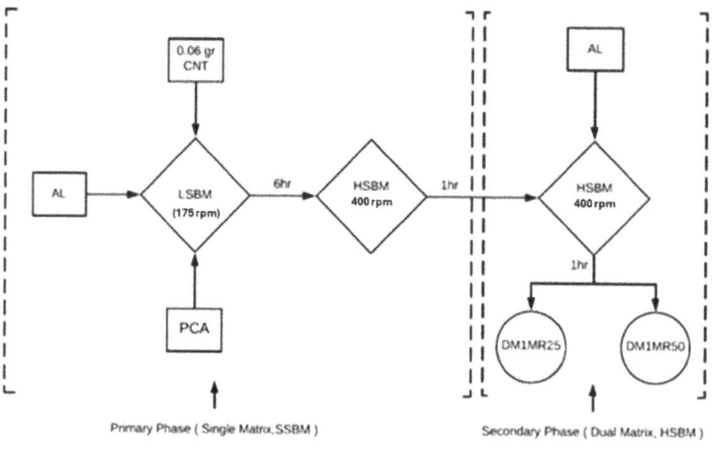

Figure 2. The AL/CNT composite powders milling steps flowchart.

The DM-SSBM composites were mixed by a planetary ball-milling machine. The CNT powders were first dispersed in the Al matrix using the SSBM method. The powders were milled in a 250 mL stainless steel jar with the BPR = 5:1 at a speed of 135 rpm for 6 h (LSBM) and 270 rpm for 1 h (HSBM) at room temperature [45]. The milling process had 5-minute stops every 10 min to prevent excessive temperature increases. A 1 wt% solution of stearic acid was used as a processing control agent (PCA) to prevent excessive cold welding between Al particles. Figure 3 presents the equipment used for the milling process. At the second step, the dual-matrix method was applied by adding the remaining soft Al to the mixture. Then, the powders were milled for an optimum time of 60 min [15] to obtain a better dispersion ratio. Argon gas was used during the milling to prevent the powder from oxidation. The DM phase ratio studied in this paper was 25% and 50% of reinforced to unreinforced Al. After the milling step. The naming system in this study will be as follows:

$$DM_xMR_{yy}\text{-}SSBM$$

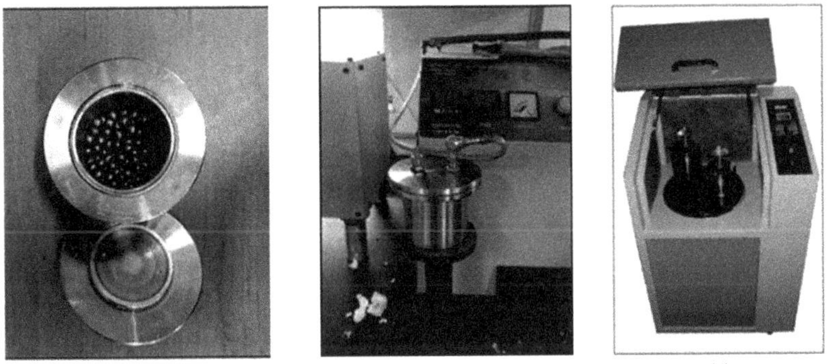

(a) (b) (c)

Figure 3. (**a**) The stainless steel milling jar filled with the milling balls, (**b**) controlled Argon insertion procedure, (**c**) the planetary ball-milling machine.

The x subscript refers to the weight percentage of the CNTs. The y subscripts refer to the overall mixing ratio of the reinforced Al matrix to the unreinforced particles.

Three other samples using HSBM, SSBM, and DM were synthesized in order to compare the final mechanical properties. The HSBM powder was milled in the same jar under the same conditions for 5 h at 400 rpm. DM and SSBM powders were synthesized under the same conditions as [14,15]. Key process parameters (milling speed, time, BPR, dual-matrix ratios) were selected based on prior reports, demonstrating their critical influence on CNT dispersion and mechanical outcomes. These parameter choices were designed to optimize the balance between sufficient UTS and ductility by preservation of CNT integrity and controlling the amount of cold welding.

2.3. Consolidation of Al/CNT Powders

All powders were consolidated as rectangular planar parts using a stainless steel die according to Figure 4, with a 65 mm width and 5 mm thickness, using a 300 Mpa uniaxial press for 1 h. After pressing, the samples were sintered at a temperature of 550 °C for 5 h using a controlled atmosphere to prevent any oxidation. In order to gain better dispersion and higher density, the sintered samples were hot-rolled at 480 °C 17 times to reach 1.6 mm thickness. The reduction percentage for each rolling pass is 4% (0.2 mm). After each set of 4 passes, due to temperature loss, samples were placed in a furnace and heated up to 480 °C again to perform flawless rolling. During the process, all the samples were covered by aluminum foil to prevent contamination and excessive temperature loss. The same consolidation method was used for all the different samples. Although CNTs are initially dispersed onto individual aluminum powder particles, the subsequent high-speed ball milling, sintering, and rolling processes result in a continuous aluminum matrix in the final composite, which physically encapsulates and interacts with the CNTs.

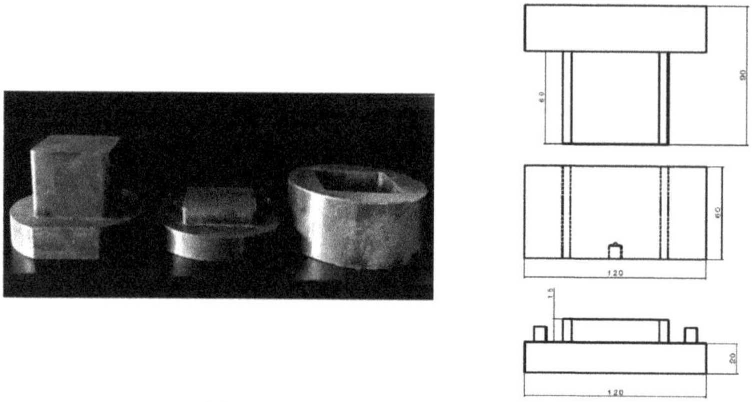

(a) (b)

Figure 4. (**a**) The stainless steel press die, (**b**) schematic of the die.

3. Results and Discussion

3.1. Al/CNT Powder Morphology

There are three different phases in the milling process to obtain the final DM-SSBM powder. Figure 5 illustrates the overall dispersion mechanism of the CNT powders during the PBM process. In Figure 5a, under the LSBM process, CNT clusters are dispersed on the Al flake powder surface, while the powder flattening is happening due to the collision between milling balls and the spherical Al powders. At this stage, no cold welding is happening between the Al particles, due to the low colliding energy and presence of the PCA (stearic acid). Figure 5b presents the HSBM phase, where the flattened Al particles containing CNTs on the surface are joining together and forming bigger particles due to the cold welding happening under high-energy collision between milling balls. During the last phase, pure aluminum is added to the mixed powder through an additional step of HSBM, applying a higher rotational speed to increase the cold welding. This approach ensures that the final product remains ductile while enhancing its yield strength.

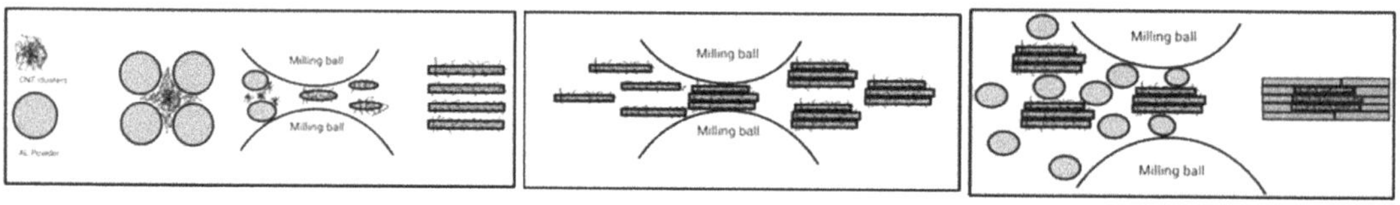

(a) (b) (c)

Figure 5. Dispersion and deformation/welding mechanism in (**a**) LSBM, (**b**) HSBM, and (**c**) DM-SSBM of Al/CNT powders.

Scanning electron microscopy (SEM) analyses of both powder and bulk Al/CNT composites were performed using a TESCAN MIRA3 XMU. As shown in Figure 6, the difference between powder morphology at various stages is clearly visible. As mentioned in the previous paragraph, after the LSBM stage, Al particles become flattened, forming a flake powder shape, with CNT particles well dispersed on the flattened surface, as shown in Figure 6a. In Figure 6b, after the first HSBM stage, particle size increases due to the cold welding between the flake AL powders. The average particle size is increased significantly to the range of 40–60 μm. After the first HSBM process, to obtain the DM phase, some pure AL powders were added to the mixture, and the process went through another HSBM process. Figure 6c illustrates the size and shape difference between the milled and unmilled AL powders, with bigger particles being the milled and welded Al particles.

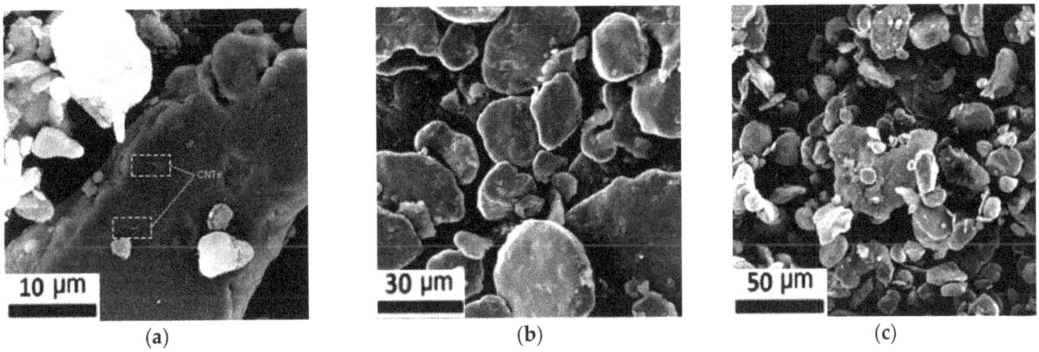

Figure 6. FE-SEM: (**a**) CNT dispersion on the Al particles after the LSBM, (**b**) increased flake powder dimension after HSBM, (**c**) comparing the pure Al particles (smaller round particles) with the milled ones (bigger flattened particles) in the DM phase.

Figure 7 analyzes the dispersion quality of the CNTs on Al particles after the LSBM phase. As shown, the dispersion quality is acceptable with some small clusters observed on certain flake surfaces, as shown in Figure 7b, and no significant clusters are present between the Al particles.

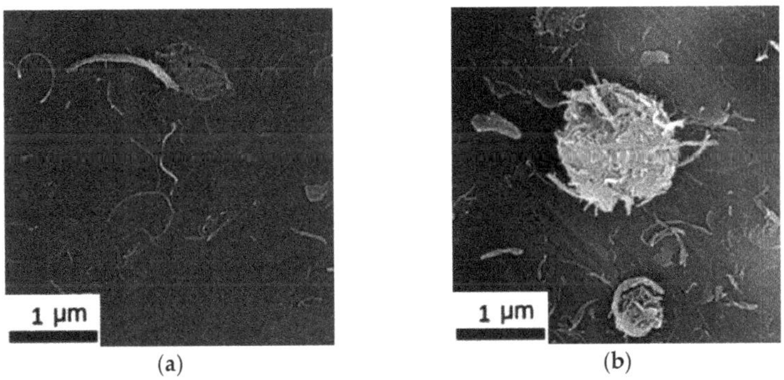

Figure 7. (**a**) CNT dispersion on the Al powder surface, (**b**) small CNT cluster found on Al powder surface.

Figure 8 presents SEM images of the Al/CNT composite powder after the final DM-SSBM stage. As it is shown in Figure 8a, after conducting HSBM and adding pure Al to the milled powder mixture, no CNT clusters are visible. Investigating the particle surfaces revealed some CNT on the powder surface (Figure 8b,c). Their presence on the powder surface confirms that the HSBM process effectively separated the agglomerates without eliminating or damaging the CNTs. This observation aligns with XRD analysis, which verifies that the CNT structure remains intact. Overall, these findings demonstrate that the HSBM phase significantly improves CNT dispersion by eliminating agglomerates and promoting homogeneous integration into the aluminum matrix, without compromising the structural integrity of the CNTs.

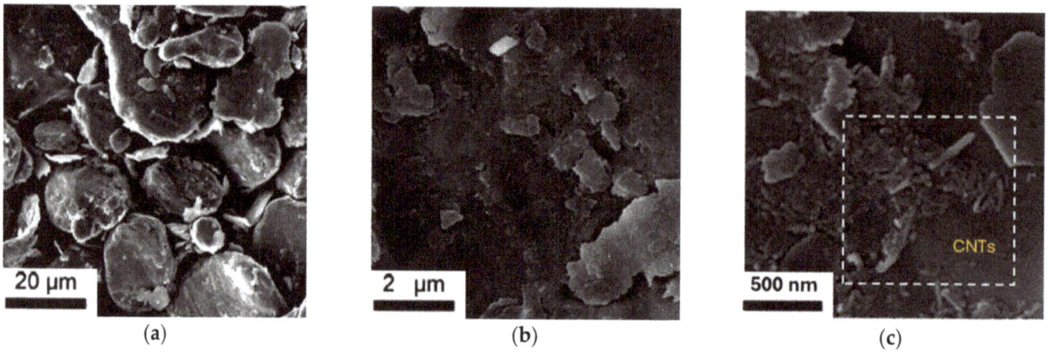

Figure 8. SEM images of final DM-SSBM sample: (**a**) 20 μm, (**b**) 2 μm, and (**c**) 500 nm.

X-ray diffraction analysis was performed following the θ–2θ method, using a Bruker Advance D8 X-ray diffractometer (XRD). In this configuration, the X-ray tube remains stationary while the sample and detector rotate at speeds of θ°/min and 2θ°/min, respectively. The scan range was set between 5° and 80°, with a step size of 0.04°. The first and most intense peak appears at 38°, corresponding to the {111} plane of aluminum. Subsequent peaks are observed at 46°, 66°, and 79°, which correspond to the {200}, {220}, and {311} planes, respectively. This diffraction pattern is in full agreement with the standard reference data for pure aluminum powder available in the PDF2 database. Based on this, it can be concluded that in samples produced with DM, SSBM, and DM-SSBM methods, no peaks associated with aluminum carbide (Al_4C_3) are present, indicating that this phase was not formed during the milling process [28].

In contrast, the HSBM sample exhibits a weak diffraction peak at 69°, which corresponds to aluminum carbide (Al_4C_3). Although this peak appears with relatively low intensity, it confirms the presence of Al_4C_3 in the matrix. These results suggest that prolonged high-energy ball milling (HSBM) leads to the inevitable formation of the aluminum carbide phase, which could negatively affect the final properties of the composite material. Figure 9 shows the XRD analysis graphs for both samples.

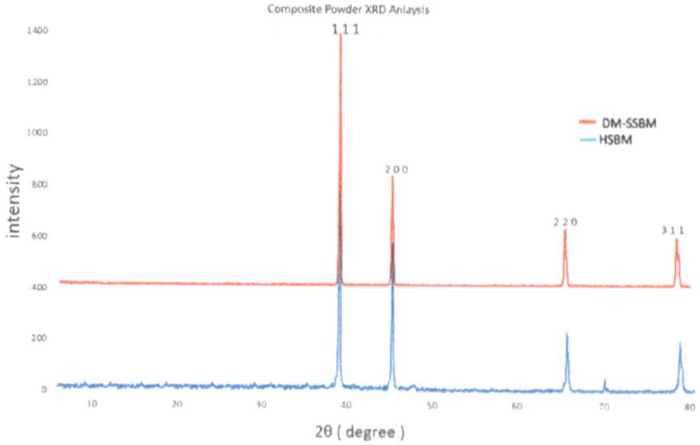

Figure 9. XRD analysis of powder samples prepared by DM-SSBMHSBM methods.

The microstructural and phase analysis results confirm that the DM-SSBM powder preparation method results in uniform CNT dispersion with no formation of the aluminum carbide (Al_4C_3) phases. The sequential low- and high-speed milling, followed by adding pure Al, balances the cold welding and prevents damaging the CNT structure, compared to the conventional HSBM approach.

3.2. Post-Rolling Microstructural Analysis

Figure 10 shows SEM images of the DM1MR50-SSBM composite after hot rolling. The surface displays long, plate-like features that run parallel to the rolling direction, confirming that the material was strongly compressed and its grains were stretched out during rolling. Importantly, no carbon-nanotube (CNT) fragments can be seen on the surface; only the aluminum matrix is visible. Given the cold-welding, sintering, and hot-rolling stages, the final bulk composite no longer consists of isolated particles, but rather a continuous metallic matrix embedding the nanotubes. This observation justifies referring to the aluminum phase as a matrix. Two factors likely explain why CNTs no longer appear at the surface:

- High rolling pressure pushes any surface CNTs slightly below the top layer as the metal flows.
- The elevated rolling temperature softens the aluminum, allowing it to wrap around and fully cover the CNTs.

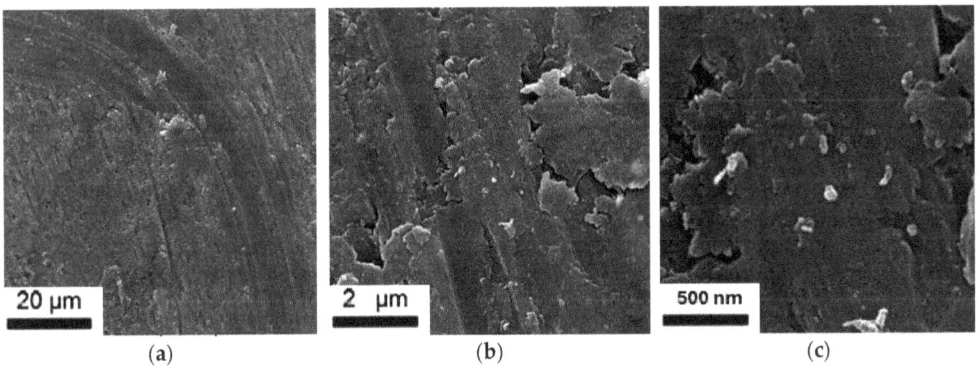

Figure 10. SEM micrographs of the DM1MR50-SSBM composite after hot rolling, showing rolling-induced deformation texture in (**a**) 20 µm, (**b**) 2 µm, and (**c**) 500 nm.

Because SEM detects only the surface of the samples, CNTs buried just beneath the surface are invisible in these images. This is beneficial as it prevents them from pulling out at the surface, and positions them inside the load-bearing aluminum, helping to explain the good toughness reported in Section 3.2.

3.3. Mechanical Properties

To evaluate the mechanical properties of the Al/CNT composites, tensile tests using ASTM B557M-15 [50] standard were conducted on all samples, including those produced via HSBM, SSBM, DM, and the combined DM-SSBM. The reported tensile strength values represent the mean of three repeated measurements, conducted in accordance with ISO/IEC 17025 [51] standards. Figure 11 shows the geometry of the test samples. According to the ASTM standard, the test was conducted at room temperature (25 °C) and 40% humidity. Three specimens per composition were tested to ensure repeatability, and the average values were reported.

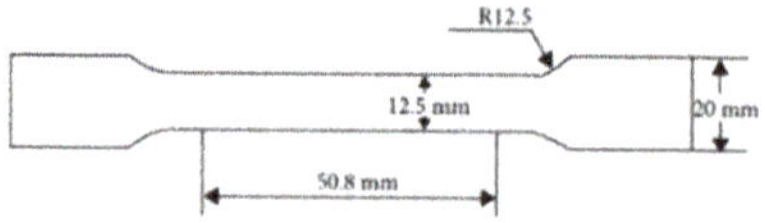

Figure 11. ASTM B557M-15 sample dimensions.

Results for the mechanical tests are presented in Figure 12.

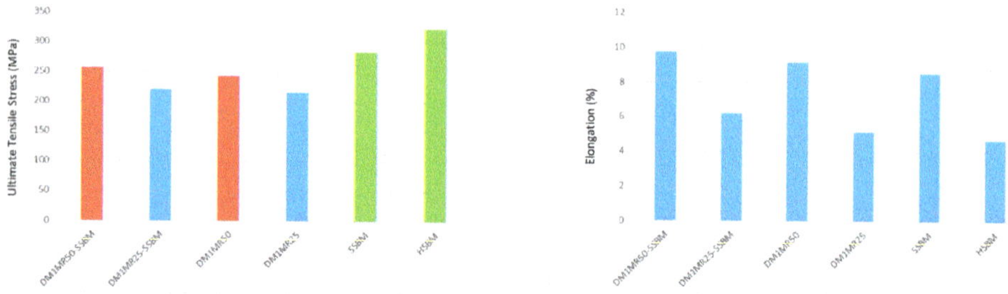

(a) (b)

Figure 12. (**a**) Ultimate tensile strength (UTS) comparison among different samples, showing the highest UTS in SSBM and HSBM (Red for MR50, Blue for MR25 and Green for other 2 methods). (**b**) Elongation (%) comparison among the same sample groups.

As depicted in Figure 12, adding the shift-speed ball-milling (SSBM) step to the dual-matrix (DM) powders gives an acceptable raise to the final UTS in the manufactured composite, and, at the 50% mixing ratio (DM_1MR_{50}), the SSBM-combined sample reaches roughly 267 MPa—around 3% higher than the DM-only sample—as shown in the ultimate tensile strength bar chart. The same pattern appears at 25%: DM1MR25-SSBM is stronger than DM_1MR_{25}. This trend is due to the high-energy milling stage in SSBM, which increases cold-welding between flakes and disperses the carbon nanotubes more uniformly, improving load transfer without introducing additional carbide form into the matrix. Meanwhile, the unmilled AL that was added in the dual-matrix stage limits overwork and keeps residual stresses low. As a result, the DM-SSBM samples sit between the aggressive HSBM and the softer DM powders in terms of ultimate stress. This trend is clearly visible in both the stress–strain curve and the UTS bar chart, where the HSBM sample achieves 328 MPa and the DM-only routes are closer to 245–260 MPa. The steep rise and sudden drop of the HSBM curve (blue) in the stress–strain plot also emphasize its brittle nature and strong cold-welding response.

The DM-SSBM compositions also deliver more ductility than either the single-stage SSBM or the high-energy HSBM samples. DM1MR50-SSBM has about 13% tensile strain, whereas SSBM levels off near 11% and HSBM fails below 5. These trends are clearly visible by the stress–strain curves depicted in Figure 13: the DM_1MR_{50}-SSBM extends furthest along the x-axis, while the blue HSBM curve fails quickly after peak stress. The softer, unmilled Al that defines the dual-matrix design provides local zones that can yield and blunt micro-cracks once the CNT-reinforced areas have carried the peak load. Meanwhile, the low-speed phase of SSBM limits CNT breakage and controls the Al_4C_3 formation, so the composite preserves its ability to stretch.

Additionally, the tensile test results illustrated in Figure 13 reveal a considerable difference in energy absorption capability of the samples. The area under the stress-strain curve, which represents the mechanical energy absorbed per unit volume before failure, is a critical indicator of the material's toughness. The DM_1MR_{50}-SSBM sample manufactured by the new proposed method in this study absorbed approximately 23.4 MJ/m^3, which is more than SSBM and DM_1MR_{50} with 21.8 MJ/m^3 and 21.2 MJ/m^3, respectively. The higher toughness in the DM-SSBM sample comes from its combination of higher ultimate tensile strength and greater elongation, which together produce a larger area under the stress-strain curve. As a result, DM_1MR_{50}-SSBM offers the most favorable mechanical profile among the tested samples, making it an excellent candidate for applications requiring materials that can absorb impact energy efficiently, such as lightweight crash components, protective barriers, or high-performance structural parts. Table 2 summarizes the information about energy absorption between the three samples.

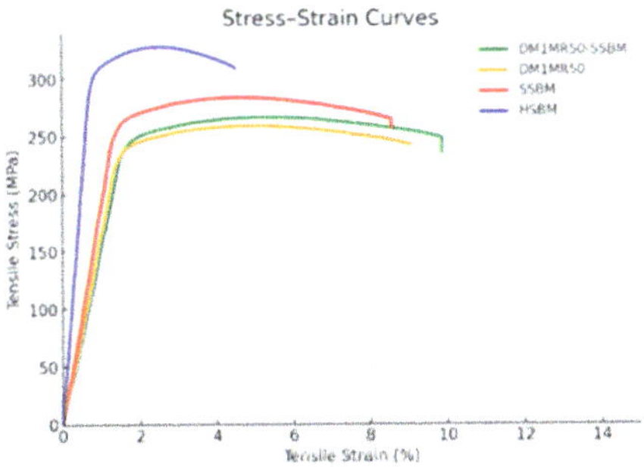

Figure 13. Stress–strain curves for DM1MR50-SSBM, DM1MR50, SSBM, and HSBM samples.

Table 2. Mechanical properties of DM1MR50, DM1MR50-SSBM, and SSBM samples.

Property	DM1MR50	DM1MR50-SSBM	SSBM
Ultimate Tensile Strength (MPa)	259.5	266.7	284
Elongation (%)	9.2	9.9	8.6
Energy Absorption (MJ/m^3)	21.2	23.4	21.8
Relative Energy Gain (%)	Baseline	+10%	+3%

Overall, the combined mechanical analysis clearly demonstrates that the DM1MR50-SSBM sample delivers the most advantageous balance of strength, ductility, and energy absorption among the tested composites. These improvements are directly linked to its optimized processing route, which combines dual-matrix design with shift-speed ball milling. While additional microstructural characterization (e.g., fracture surface analysis) could provide deeper insights, the current findings already highlight the strong potential of the DM-SSBM route for applications demanding high toughness and reliable energy dissipation.

4. Conclusions

In this study, a novel flake powder metallurgy approach combining dual-matrix (DM) design with shift-speed ball milling (SSBM) was successfully applied to fabricate Al/CNT composites with enhanced mechanical properties. Compared to conventional HSBM, SSBM, and DM methods, the DM-SSBM method delivered the most balanced performance, achieving both high ultimate tensile strength (~267 MPa) and superior ductility (~9.9%). Tensile testing revealed that DM-SSBM samples absorbed approximately 23.4 MJ/m^3 of mechanical energy before failure—outperforming both the DM_1MR_{50} (21.2 MJ/m^3) and SSBM (21.8 MJ/m^3). This improvement comes from the optimized combination of CNT dispersion, cold welding, and controlled matrix softening, which together enhance both load-bearing capacity and plastic deformation ability.

Furthermore, phase analysis confirmed that the DM-SSBM method effectively prevented Al_4C_3 formation, preserving the integrity of the CNT reinforcement. Although simulation and modeling approaches offer powerful tools for many composite systems, accurately modeling nanoscale interactions between CNTs and the aluminum matrix remains nearly impossible with current computational resources. Therefore, this study relies on experimentally validated methods, supported by literature, to achieve systematic material optimization.

These findings highlight the potential of DM-SSBM composites for applications requiring lightweight, high-toughness materials—particularly in energy-absorbing components such as crash structures, protective shells, and advanced structural parts. Future work should explore the microstructural evolution under cyclic loading and evaluate long-term durability to fully map the industrial potential of these advanced composites.

Author Contributions: Conceptualization, H.R.; methodology, H.R.; investigation, H.R.; resources, H.R.; writing—original draft preparation, H.R., writing—review and editing, A.V.; supervision, A.V. All authors have read and agreed to the published version of the manuscript.

Funding: This research received no external funding.

Data Availability Statement: The raw data supporting the conclusions of this article will be made available by the authors on request.

Conflicts of Interest: The authors declare no conflicts of interest.

Abbreviations

The following abbreviations are used in this manuscript:

CNT	Carbon Nanotubes
MWCNT	Multi-wall Carbon Nanotubes
SWCNT	Single-wall Carbon Nanotubes
BPR	Ball-to-Powder Ratio
LSBM	Low-speed Ball Milling
HSBM	High-speed Ball Milling
SSBM	Shift-speed Ball Milling
DM	Dual Matrix
DM-SSBM	Dual-matrix Shift-speed Ball Milling
MR	Mixing Ratio

References

1. Khanna, V.; Kumar, V.; Bansal, S.A. Mechanical properties of aluminium-graphene/carbon nanotubes (CNTs) metal matrix composites: Advancement, opportunities and perspective. *Mater. Res. Bull.* **2021**, *138*, 111224. [CrossRef]
2. Ogawa, F.; Masuda, C. Fabrication and the mechanical and physical properties of nanocarbon-reinforced light metal matrix composites: A review and future directions. *Mater. Sci. Eng. A* **2021**, *820*, 141542. [CrossRef]
3. Cha, S.I.; Kim, K.T.; Arshad, S.N.; Mo, C.B.; Hong, S.H. Extraordinary strengthening effect of carbon nanotubes in metal-matrix nanocomposites processed by molecular-level mixing. *Adv. Mater.* **2005**, *17*, 1377–1381. [CrossRef] [PubMed]
4. Dong, S.; Tu, J.; Zhang, X. An investigation of the sliding wear behavior of Cu-matrix composite reinforced by carbon nanotubes. *Mater. Sci. Eng. A* **2001**, *313*, 83–87. [CrossRef]
5. Agarwal, A.; Bakshi, S.R.; Lahiri, D. *Carbon Nanotubes: Reinforced Metal Matrix Composites*; CRC press: Boca Raton, FL, USA, 2018.
6. George, R.; Kashyap, K.; Rahul, R.; Yamdagni, S. Strengthening in carbon nanotube/aluminium (CNT/Al) composites. *Scr. Mater.* **2005**, *53*, 1159–1163. [CrossRef]
7. Curtin, W.A.; Sheldon, B.W. CNT-reinforced ceramics and metals. *Mater. Today* **2004**, *7*, 44–49. [CrossRef]

8. Seidel, G.D.; Lagoudas, D.C. Micromechanical analysis of the effective elastic properties of carbon nanotube reinforced composites. *Mech. Mater.* **2006**, *38*, 884–907. [CrossRef]
9. Coleman, J.N.; Khan, U.; Blau, W.J.; Gun'ko, Y.K. Small but strong: A review of the mechanical properties of carbon nanotube–polymer composites. *Carbon* **2006**, *44*, 1624–1652. [CrossRef]
10. Ajayan, P.M.; Schadler, L.S.; Giannaris, C.; Rubio, A. Single-walled carbon nanotube–polymer composites: Strength; weakness. *Adv. Mater.* **2000**, *12*, 750–753. [CrossRef]
11. Bakshi, S.R.; Lahiri, D.; Agarwal, A. Carbon nanotube reinforced metal matrix composites-a review. *Int. Mater. Rev.* **2010**, *55*, 41–64. [CrossRef]
12. Saheb, N.; Iqbal, Z.; Khalil, A.; Hakeem, A.S.; Al Aqeeli, N.; Laoui, T.; Al-Qutub, A.; Kirchner, R.; Francis, L.D. Spark plasma sintering of metals and metal matrix nanocomposites: A review. *J. Nanomater.* **2012**, *2012*, 18. [CrossRef]
13. Neubauer, E.; Kitzmantel, M.; Hulman, M.; Angerer, P. Potential and challenges of metal-matrix-composites reinforced with carbon nanofibers and carbon nanotubes. *Compos. Sci. Technol.* **2010**, *70*, 2228–2236. [CrossRef]
14. Xu, R.; Tan, Z.; Xiong, D.; Fan, G.; Guo, Q.; Zhang, J.; Su, Y.; Li, Z.; Zhang, D. Balanced strength and ductility in CNT/Al composites achieved by flake powder metallurgy via shift-speed ball milling. *Compos. Part A Appl. Sci. Manuf.* **2017**, *96*, 57–66. [CrossRef]
15. Salama, E.I.; Abbas, A.; Esawi, A.M. Preparation and properties of dual-matrix carbon nanotube-reinforced aluminum composites. *Compos. Part A Appl. Sci. Manuf.* **2017**, *99*, 84–93. [CrossRef]
16. Rikhtegar, F.; Shabestari, S.; Saghafian, H. Microstructural evaluation and mechanical properties of Al-CNT nanocomposites produced by different processing methods. *J. Alloys Compd.* **2017**, *723*, 633–641. [CrossRef]
17. Liu, Z.; Xiao, B.; Wang, W.; Ma, Z. Modelling of carbon nanotube dispersion and strengthening mechanisms in Al matrix composites prepared by high energy ball milling-powder metallurgy method. *Compos. Part A Appl. Sci. Manuf.* **2017**, *94*, 189–198. [CrossRef]
18. Liu, X.; Li, C.; Eckert, J.; Prashanth, K.G.; Renk, O.; Teng, L.; Liu, Y.; Bao, R.; Tao, J.; Shen, T.; et al. Microstructure evolution and mechanical properties of carbon nanotubes reinforced Al matrix composites. *Mater. Charact.* **2017**, *133*, 122–132. [CrossRef]
19. Huang, H.; Fan, G.; Tan, Z.; Xiong, D.-B.; Guo, Q.; Guo, C.; Li, Z.; Zhang, D. Superplastic behavior of carbon nanotube reinforced aluminum composites fabricated by flake powder metallurgy. *Mater. Sci. Eng. A* **2017**, *699*, 55–61. [CrossRef]
20. Chen, B.; Shen, J.; Ye, X.; Jia, L.; Li, S.; Umeda, J.; Takahashi, M.; Kondoh, K. Length effect of carbon nanotubes on the strengthening mechanisms in metal matrix composites. *Acta Mater.* **2017**, *140*, 317–325. [CrossRef]
21. Zare, H.; Jahedi, M.; Toroghinejad, M.R.; Meratian, M.; Knezevic, M. ; Compressive; shear, and fracture behavior of CNT reinforced Al matrix composites manufactured by severe plastic deformation. *Mater. Des.* **2016**, *106*, 112–119. [CrossRef]
22. Mokdad, F.; Chen, D.; Liu, Z.; Xiao, B.; Ni, D.; Ma, Z. Deformation and strengthening mechanisms of a carbon nanotube reinforced aluminum composite. *Carbon* **2016**, *104*, 64–77. [CrossRef]
23. Liu, Z.; Zhao, K.; Xiao, B.; Wang, W.; Ma, Z. Fabrication of CNT/Al composites with low damage to CNTs by a novel solution-assisted wet mixing combined with powder metallurgy processing. *Mater. Des.* **2016**, *97*, 424–430. [CrossRef]
24. Chen, B.; Kondoh, K.; Imai, H.; Umeda, J.; Takahashi, M. Simultaneously enhancing strength and ductility of carbon nanotube/aluminum composites by improving bonding conditions. *Scr. Mater.* **2016**, *113*, 158–162. [CrossRef]
25. Chen, B.; Kondoh, K.; Imai, H.; Umeda, J. Effect of initial state on dispersion evolution of carbon nanotubes in aluminium matrix composites during a high-energy ball milling process. *Powder Metall.* **2016**, *59*, 216–222. [CrossRef]
26. Carvalho, O.; Miranda, G.; Soares, D.; Silva, F. Carbon nanotube dispersion in aluminum matrix composites—Quantification and influence on strength. *Mech. Adv. Mater. Struct.* **2016**, *23*, 66–73. [CrossRef]
27. Simões, S.; Viana, F.; Reis, M.A.; Vieira, M.F. Influence of dispersion/mixture time on mechanical properties of Al–CNTs nanocomposites. *Compos. Struct.* **2015**, *126*, 114–122. [CrossRef]
28. Liao, J.; Tan, M.-J. Mixing of carbon nanotubes (CNTs) and aluminum powder for powder metallurgy use. *Powder Technol.* **2011**, *208*, 42–48. [CrossRef]
29. Jiang, L.; Li, Z.; Fan, G.; Cao, L.; Zhang, D. The use of flake powder metallurgy to produce carbon nanotube (CNT)/aluminum composites with a homogenous CNT distribution. *Carbon* **2012**, *50*, 1993–1998. [CrossRef]
30. Liu, Z.; Xu, S.; Xiao, B.; Xue, P.; Wang, W.; Ma, Z. Effect of ball-milling time on mechanical properties of carbon nanotubes reinforced aluminum matrix composites. *Compos. Part A Appl. Sci. Manuf.* **2012**, *43*, 2161–2168. [CrossRef]
31. Kwon, H.; Leparoux, M. Hot extruded carbon nanotube reinforced aluminum matrix composite materials. *Nanotechnology* **2012**, *23*, 415701. [CrossRef]
32. Kim, K.T.; Cha, S.I.; Hong, S.H. Hardness and wear resistance of carbon nanotube reinforced Cu matrix nanocomposites. *Mater. Sci. Eng. A* **2007**, *449*, 46–50. [CrossRef]

33. Lin, C.; Chang, Z.-C.; Tung, Y.; Ko, Y.-Y. Manufacturing and tribological properties of copper matrix/carbon nanotubes composites. *Wear* **2011**, *270*, 382–394. [CrossRef]
34. Shukla, A.; Nayan, N.; Murty, S.; Mondal, K.; Sharma, S.; George, K.M.; Bakshi, S.R. Processing copper–carbon nanotube composite powders by high energy milling. *Mater. Charact.* **2013**, *84*, 58–66. [CrossRef]
35. Chen, X.; Li, W.; Chen, C.; Xu, L.; Yang, Z.; Hu, J. Preparation and properties of Cu matrix composite reinforced by carbon nanotubes. *Trans. Nonferr. Met. Soc. China* **2005**, *15*, 314–318.
36. Arribas, A.S.; Bermejo, E.; Chicharro, M.; Zapardiel, A.; Luque, G.L.; Ferreyra, N.F.; Rivas, G.A. Analytical applications of a carbon nanotubes composite modified with copper microparticles as detector in flow systems. *Anal. Chim. Acta* **2006**, *577*, 183–189. [CrossRef]
37. Chen, D.; Chen, L.; Liu, S.; Ma, C.; Chen, D.; Wang, L. Microstructure and hydrogen storage property of Mg/MWNTs composites. *J. Alloys Compd.* **2004**, *372*, 231–237. [CrossRef]
38. Carreño-Morelli, E.; Yang, J.; Couteau, E.; Hernadi, K.; Seo, J.W.; Bonjour, C.; Forró, L.; Schaller, R. Carbon nanotube/magnesium composites. *Phys. Status Solidi (A)* **2004**, *201*, R53–R55. [CrossRef]
39. Kondoh, K.; Fukuda, H.; Umeda, J.; Imai, H.; Fugetsu, B.; Endo, M. Microstructural and mechanical analysis of carbon nanotube reinforced magnesium alloy powder composites. *Mater. Sci. Eng. A* **2010**, *527*, 4103–4108. [CrossRef]
40. Li, C.; Wang, X.; Wu, K.; Liu, W.; Xiang, S.; Ding, C.; Hu, X.; Zheng, M. Distribution and integrity of carbon nanotubes in carbon nanotube/magnesium composites. *J. Alloys Compd.* **2014**, *612*, 330–336. [CrossRef]
41. He, C.; Zhao, N.; Shi, C.; Du, X.; Li, J.; Li, H.; Cui, Q. An approach to obtaining homogeneously dispersed carbon nanotubes in Al powders for preparing reinforced Al-matrix composites. *Adv. Mater.* **2007**, *19*, 1128–1132. [CrossRef]
42. Zhou, W.; Bang, S.; Kurita, H.; Miyazaki, T.; Fan, Y.; Kawasaki, A. Interface and interfacial reactions in multi-walled carbon nanotube-reinforced aluminum matrix composites. *Carbon* **2016**, *96*, 919–928. [CrossRef]
43. Rikhtegar, F.; Shabestari, S.; Saghafian, H. The homogenizing of carbon nanotube dispersion in aluminium matrix nanocomposite using flake powder metallurgy and ball milling methods. *Powder Technol.* **2015**, *280*, 26–34. [CrossRef]
44. Li, Q.; Rottmair, C.A.; Singer, R.F. CNT reinforced light metal composites produced by melt stirring and by high pressure die casting. *Compos. Sci. Technol.* **2010**, *70*, 2242–2247. [CrossRef]
45. Chen, M.; Fan, G.; Tan, Z.; Xiong, D.; Guo, Q.; Su, Y.; Zhang, J.; Li, Z.; Naito, M.; Zhang, D. Design of an efficient flake powder metallurgy route to fabricate CNT/6061Al composites. *Mater. Des.* **2018**, *142*, 288–296. [CrossRef]
46. Choi, H.; Shin, J.; Bae, D. The effect of milling conditions on microstructures and mechanical properties of Al/MWCNT composites. *Compos. Part A Appl. Sci. Manuf.* **2012**, *43*, 1061–1072. [CrossRef]
47. Morsi, K.; Esawi, A. Effect of mechanical alloying time and carbon nanotube (CNT) content on the evolution of aluminum (Al)–CNT composite powders. *J. Mater. Sci.* **2007**, *42*, 4954–4959. [CrossRef]
48. Bakshi, S.R.; Agarwal, A. An analysis of the factors affecting strengthening in carbon nanotube reinforced aluminum composites. *Carbon* **2011**, *49*, 533–544. [CrossRef]
49. Esawi, A.M.; El Borady, M.A. Carbon nanotube-reinforced aluminium strips. *Compos. Sci. Technol.* **2008**, *68*, 486–492. [CrossRef]
50. *ASTM B557M-15*; Standard Test Methods for Tension Testing Wrought and Cast Aluminum- and Magnesium-Alloy Products (Metric). ASTM: West Conshohocken, PA, USA, 2023. Available online: https://store.astm.org/b0557m-15.html (accessed on 1 June 2025).
51. *ISO/IEC 17025*; Testing and Calibration Laboratories. ISO: Geneva, Switzerland, 2017. Available online: https://www.iso.org/publication/PUB100424.html (accessed on 1 June 2025).

MDPI AG
Grosspeteranlage 5
4052 Basel
Switzerland
Tel.: +41 61 683 77 34

Designs Editorial Office
E-mail: designs@mdpi.com
www.mdpi.com/journal/designs

Disclaimer/Publisher's Note: The title and front matter of this reprint are at the discretion of the Guest Editor. The publisher is not responsible for their content or any associated concerns. The statements, opinions and data contained in all individual articles are solely those of the individual Editor and contributors and not of MDPI. MDPI disclaims responsibility for any injury to people or property resulting from any ideas, methods, instructions or products referred to in the content.

www.ingramcontent.com/pod-product-compliance
Lightning Source LLC
Chambersburg PA
CBHW061621210926
51850CB00009B/2399